This book is everything I love… endorses my feeling and mantra in life that health is wealth and that we are able with great people guiding like Nicole – to really get in charge of our own optimal health. To put the pieces of the jigsaw together and achieve the healthiest version of ourselves. This book is easy to read and understand, makes sense, and has so much valuable information. Nicole herself is compassionate, passionate and has put together a well thought out programme – I highly recommend this book.

Emma Forbes – TV and Radio Presenter

A *must* for anyone experiencing autoimmunity, gut health disorders, burnout and much more. Nicole's approach to addressing disease and optimising health and longevity is holistic and thorough. She breaks down the complex web of health in a way that will leave you awe-inspired and ready to take action!

Dr Lafina Diamandis – GP and Lifestyle Medicine Doctor

I loved reading this book. This is exactly the kind of book I wish I had read at the start of my own journey to optimizing my health as an entrepreneur. It's full of so much valuable information and lots of helpful tips to start prioritizing your health. It's great to see another practitioner taking a holistic approach and dealing with the root causes of an individual's symptoms. I highly recommend this book for anyone looking to optimize their health and energy so they can perform at their best.

Lorna Devine – Psychologist, Cognitive Behavioural Therapist and High-Performance Coach

Brilliant book. If you want to understand your health better and what you can do to support optimal energy, immunity and longevity this is an absolute must read!

Farzanah Nasser – Functional Medicine Practitioner

Nicole is the real deal. As a serial entrepreneur, working with her to overcome my chronic fatigue and to finally optimise my health and energy levels has been a game-changer! If you want to feel better, do better and LIVE better, her book should be a non-negotiable on your nightstand!

Chris Ducker – Bestselling author of 'Rise of the Youpreneur'

Debussy said that, 'music is the space between the notes'. When you're seeking to optimise your health, there is a lot of crash-bang-wallop out there. Nicole's MitoImmune Method offers you the 'Why?' and – importantly – simple, bite-sized actions you can take. The result? Space to be able to hear once again the melody of your optimal health. With Nicole's help, I've transformed my autoimmune conditions.

Eric Ho – Executive and Health Coach, Founder of Health for Success

Looking after your health can be confusing and overwhelming. But with Nicole's MitoImmune Method, I've been able to transform my energy and immunity. I can't recommend this book highly enough!

Will Higgins – Founder Nu Mind and Wellness

Working with Nicole has completely changed my life! Before, I was a 26 year old unable to find the energy to complete daily tasks, let alone socialise! Six months later and I feel like a whole new person – even playing hockey on a weeknight! Nicole's knowledge and continued support throughout my journey has been absolutely incredible and I cannot thank her enough for helping me understand how my body works and the nutrition I need to help it function properly. I wish I'd found Nicole and her book sooner!

Kimberley Browne – Patient

As a previous client of Nicole's, this book epitomises her knowledge and passion for optimal health and wellbeing. A comprehensive yet practical book to remind us that optimal health is within reach for everyone, and with a bit more knowledge and understanding, put in a way that anyone can understand, no matter where you are in your health journey you can get started and change your health and your life.

Brook Thompson – Patient

I couldn't wait to read this book! My general wellbeing was not in a great place when I was put in touch with Nicole. She put me at ease from our first meeting and made me realise that with her help and guidance and my commitment I could get myself back to my Optimal health.

Sally Cotterell – Patient

Nicole's book is a fascinating and informative insight into all things autoimmune. I have learnt so much about my own health. I certainly didn't understand before reading this book that many of my health issues were autoimmune related. There is no doubt that knowledge is power and with this book we can all learn so much about how to improve our health!

Jane Downing – Patient

Optimal YOU

Supercharge your ENERGY,
Strengthen your IMMUNITY

NICOLE GOODE
FUNCTIONAL MEDICINE PRACTITIONER

First published in Great Britain by Practical Inspiration Publishing, 2024

Cover Image: Matt Priestley Photography
Make Up Artist: Liz Clough
Hair Stylist: Rebecca Kuk

ISBN 978-1-78860-634-9 (hardback)
978-1-78860-635-6 (paperback)
978-1-78860-637-0 (epub)
978-1-78860-636-3 (mobi)

To Mum and Dad for everything.

None of it would have been possible without your love and support.

~ ATTRIBUTED TO THE Dalai Lama when asked what surprises him most about humanity ~

"Man! Because he sacrifices his health in order to make money. Then he sacrifices money to recuperate his health. And then he is so anxious about the future that he does not enjoy the present; the result being that he does not live in the present or the future; he lives as if he is never going to die, and then dies having never really lived."

In the resources you can access all the tools we talk about in the book, recipes to help you move through the MitoImmune Plan, supplement recommendations and more. Find all this and more at www.nicolegoodehealth.com/optimal-you-book-resources. Password: OptimalYouBook.

Contents

Part 1: The foundations of optimal health 1
Chapter 1: Introduction 3
Chapter 2: Functional medicine 15
Chapter 3: Addressing root causes.................... 23
Chapter 4: The power of personalized plans 27
Chapter 5: The MitoImmune Method.................... 31

Part 2: The pillars of optimal health 35
Chapter 6: Mitochondrial health 37
Chapter 7: Immune health 47
Chapter 8: Brain health 57
Chapter 9: Adrenal health 67
Chapter 10: Thyroid health 77
Chapter 11: Gut health 87
Chapter 12: Hormone health....................97
Chapter 13: Cardiometabolic health 107

Part 3: The MitoImmune Way – your path to optimal health....................117
Chapter 14: The MitoImmune Programme....................119
Chapter 15: Specific nutrients 139
Chapter 16: Lifestyle – the 4 S's....................161
Chapter 17: Quality of our food....................181
Chapter 18: Environmental toxins 187
Chapter 19: Functional testing....................195
Chapter 20: What next? 199

Appendix I: Resources 201

Appendix II: My favourites 203

Appendix III: References........ 207

Appendix IV: Create your personalized plan 225

Acknowledgements........ 229

About the author 231

Index........ 233

PART 1

The foundations of optimal health

1

Introduction

I'm going to put this bluntly – something is going to get you in the end. There is no miracle cure to ageing. Every day we get older whether we like it or not; we don't have the key to immortality. In fact we should see ageing as a privilege; I'm sure you all know someone who would love to still be here getting one day older.

But let's just imagine for a moment that we did have the key to immortality. *Would you want to live forever?* That's the big question isn't it? It often brings out very strong opinions in people. For some it's a hard Yes. Absolutely, who wouldn't want to live forever? They imagine themselves as immortal superhumans. For others it's a hard No. It's nature, it's the way it's supposed to be, who on earth would want to live forever?

I think that actually the question is wrong. The question isn't would you want to live forever? ***It is... if you could be well and fit would you want to live longer?***

Because for those that are hard No, the reason is always the same: you reach a point where you are old and sick and who wants to go on forever living like that. They've always imagined old age, they've imagined themselves immobile and no longer fit and healthy. And those that are hard Yes have forgotten to think about this; they've usually imagined some sci-fi scenario. I guarantee you they have never once pictured themselves living as a sick 90 year old forever.

This whole longevity picture has blown up in recent years, but as humans we don't actually want to just live longer, we want to live well for longer. Which of the following scenarios would you take:

A. You can live for 80 years fighting fit until your last day, loving and living life to the full.

   or

B. You can live for 110 years, but you will only be well for 70 of them.

I bet every single one of you would pick scenario A. So actually you aren't hunting for longevity. You are hunting for ***Optimal Health***.

As much as you can read every bit of research and every book, you cannot get away from the fact that we all age. This book isn't your miracle cure to anti-ageing (which is actually the science of looking youthful, not health – an entirely different thing). So what is this book about? To understand that let's get clear on some of the terms.

It is about extending your:

- Healthspan (the length of time you live healthily)
- Wellness medicine (staving off chronic illness for as long as possible)
- Longevity medicine (the science of living longer)
- Optimizing your health for the duration of your lifespan (how long you live)
- Optimal health (personalizing your health to maintain balance and keep you performing at the top of your game).

HEALTHSPAN VS LIFESPAN

Healthspan and lifespan are two interconnected but distinct concepts. Lifespan refers to the total number of years an individual is alive, while healthspan focuses on the years of life spent in good health and without the burden of chronic diseases. Striking a balance between the two is essential for achieving optimal wellbeing.

The importance of healthspan

Quality over quantity: Healthspan encourages a shift in perspective – from merely adding years to your life to adding life to your years. While extending lifespan is a commendable goal, it becomes more meaningful when those additional years are characterized by vibrant health, mental clarity and physical agility. Embracing healthspan as a priority ensures that the focus is on enhancing the overall quality of life.

Reducing the burden of chronic diseases: Chronic diseases, such as heart disease, diabetes, autoimmune diseases and neurodegenerative conditions, can significantly impact both lifespan and healthspan. Functional medicine delves into healthspan with the aim of helping you live well for longer by getting to the root cause. By preventing and managing chronic diseases, individuals can enjoy a longer, healthier life.

Holistic wellness: Healthspan considers the interconnectedness of various aspects of health – physical, mental and emotional. Functional medicine emphasizes the importance of a holistic approach, addressing not only symptoms, but also lifestyle factors, nutrition and stress management. This comprehensive strategy contributes to overall wellbeing, promoting healthspan by supporting the body's natural healing mechanisms.

Personalized care: Each individual is unique, with different genetic makeup, environmental exposures and lifestyle choices. Functional medicine tailors healthcare to the individual, recognizing that what works for one person may not work for another. This personalized approach ensures that interventions are targeted and effective, enhancing healthspan by addressing specific needs and vulnerabilities.

Proactive health management: Unlike traditional medicine, which often focuses on reactive measures to treat illnesses, functional medicine takes a proactive stance. By identifying risk factors early, optimizing nutrition and implementing lifestyle changes, individuals can actively manage their health and prevent the onset of chronic conditions. This proactive approach contributes to a life where you live healthier for longer – healthspan.

How this book can help you

This book is your key to ageing as well as you can, to maintaining your balance and vitality for as long as possible; it is your chance to gain the knowledge to stave off chronic illness, or to balance any chronic illness you have, so that you stay healthier for longer.

It is your key to the OPTIMAL version of YOU

To do this there are two main concepts we are going to look at and how they interconnect around the body.

Today we live in a modern and frantic world. Within this world and the need to have it all, the pursuit of optimal health, longevity and wellness becomes paramount. We all have a desire to live well and thrive, succeed in our careers, pursue our dreams, nurture our relationships and maintain physical and mental wellbeing. In this quest we can easily lose balance, but one factor to success stands out above all:

ENERGY

Energy is what makes it possible – it is the force that allows us to propel ourselves forwards. However, energy is not an infinite resource. It is finite and must be managed wisely. Like a bank account, it can be invested or depleted. Energy is part of our vitality, the currency of our productivity and the secret to unlocking our full potential. When we think about living well, supporting our health and what we need to achieve our vision of success, we think energy – the ability to have the energy we need to do all we want to do each day and to be able to work, train and spend time with family and friends.

Imagine a day when you wake up refreshed, ready to tackle your goals with enthusiasm and still have energy left to pursue your passions in the evening. That day is the day we all aspire to, but few achieve. In the world we live in today, the demands on our time and energy seem insurmountable. From demanding careers to bustling social lives, bringing up families, looking after elderly relatives and the ever-present digital distractions, we are constantly pulled in multiple directions. Our vision of success requires us to do it all, but how can we possibly have the energy to do this without burning out, without getting sick?

This brings us to the second concept:

IMMUNITY

While few manage to achieve that perfect day of vitality, some do, but those who do often end up ill; their immunity comes into the equation from a body and a mind that has been pushed to its limits. In the ever more important landscape of health and wellbeing, our immune system plays a pivotal role. Immunity is the foundation of our wellness journey – the shield that safeguards us from illness and the bridge to living a longer, healthier life.

Our immune system is a marvel of biological engineering. It is a complex network of cells, tissues and molecules working in harmony to protect us from harmful invaders. But beyond its primary function, immunity has the potential to unlock a world of resilience and longevity. It is also behind one of the fastest growing epidemics in healthcare today – autoimmune disease (more on autoimmune disease below and in chapter 7).

PEARL: Autoimmune diseases have reached epidemic levels with increases of up to 12% a year, with teenagers seeing an increase on diagnosis over the last 24 years of 300%.

In the contemporary world, our immune systems face new challenges and opportunities. Environmental stressors, dietary choices and lifestyle factors all have a profound impact on our immune function. Our daily choices can either balance or undermine immune resilience. With greater immune resilience we can push our bodies further, we won't be off work sick, we will live well for longer and be around for our families – immunity gives us longevity. Immune resilience comes down to two things: avoiding illness and living well for longer.

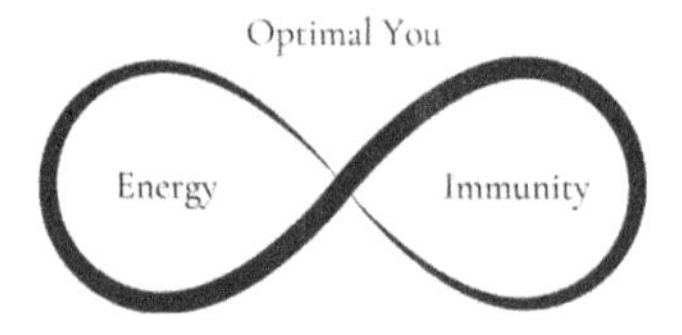

Autoimmune disease

PEARL: Once you have one autoimmune disease, the average is you will get five.

How do we stop this from happening? We deal with the root causes and balance our immune systems. This isn't said to frighten you. With one in ten now having an autoimmune diagnosis, and this figure rapidly growing, you need to know the impact on your body. Sadly conventional medicine often fails to even tell a person if the diagnosis they have is autoimmune.

I have a space on my intake form for people to write their diagnoses. Further down I have a box for them to tick: Do you have an autoimmune disease? I can't tell you how many of these intake forms I've been through where the person has written a diagnosis that is an autoimmune disease and then circled 'No' to that question. Why? Because no one has ever told them that what they have is autoimmune.

Why is this important? Because of that previous PEARL. If we don't balance our immune systems they run in overdrive and act as a risk factor, a trigger to developing another autoimmune disease. One of the most common things people say to me in our first consultation is, 'I feel like I'm collecting diseases, what is wrong with me?'

It's simple. Our environmental triggers – diet, lifestyle, toxins, infections – and our immune function drive autoimmunity. If you have an autoimmune disease you have the disease itself, but you also have autoimmunity. Let's take Hashimoto's disease, for example. Hashimoto's means you have hypothyroidism (an underactive thyroid) and you need to balance your thyroid function. This is the disease, the diagnosis, but you also have autoimmunity. Hashimoto's is an autoimmune hypothyroidism so you need to balance your immune function too. The co-occurrence – the risk of developing more autoimmune diseases once you have triggered one – is well documented.

PEARL: Autoimmune diseases are the fastest growing cause of chronic illness.

There are over 100 autoimmune diseases and one in ten people are now affected by just 19 of these – this gives you an idea of how many people are impacted by autoimmunity. We also have to consider that this is just the number of people diagnosed. Autoimmunity can be bubbling away in the body for an average of five to seven years before diagnosis. How many people do you think there are who have undiagnosed autoimmunity?

Stress is one of the most common triggers for these autoimmune conditions, with stress costing companies $1 trillion a year (more on this in a minute). Autoimmune conditions are also growing, in large part due to our environment, by 3–9% a year. This doesn't sound much but let's look at this in numbers.

In 2023, one in twenty people had a thyroid condition, and remember this is a low estimate as it doesn't include those who are not yet diagnosed. In that same year, the approximate population of the UK was 68 million. That's approximately 3.4 million people with a diagnosed thyroid condition. Now thyroid conditions have been estimated to be growing at a rate of 9–12%; they are one of the faster growing conditions. Let's take the 9%, the low end of the estimate: 9% just doesn't sound much does it? Here is where it gets interesting. Let's look at that growth year on year of people diagnosed with a thyroid condition:

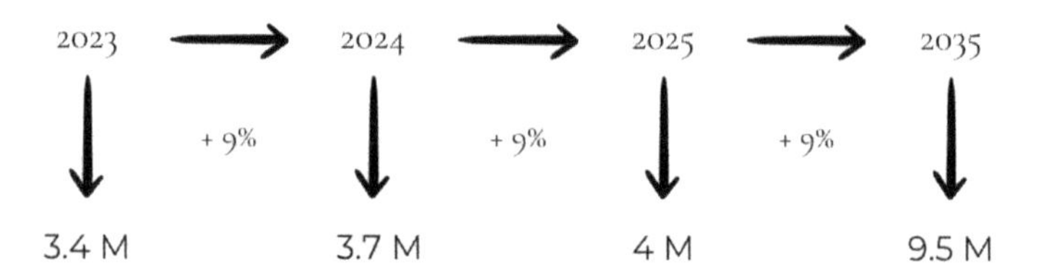

This is just for one autoimmune thyroid condition. Now factor in there are around 100 autoimmune diseases. The figures of growth are dizzying. A balanced immune system and immune resilience isn't just important; **it's vital**.

Energy (MITO) and immunity (IMMUNE)

ENERGY IMMUNITY

MITO IMMUNE

This is why I created The MitoImmune Method, focusing on energy and immunity. Mito is short for our mitochondria which, as you will see in chapter 6 when we go into cellular energy, are the powerhouses of our cells providing us with energy.

I started my journey in the work I do as an autoimmune specialist and still work heavily in this field today. It's a fact that autoimmunity is a modern world epidemic, but two things led me into working with people to optimize their health, as well as those coming from a chronic illness diagnosis.

First, the chronic stress we are under in today's society is triggering and driving autoimmunity. Almost every autoimmune patient I worked with had chronic stress as one of their triggers (there is often a group of triggers, not just one – we will look more at triggers in chapter 2). Stress was a huge problem causing an even bigger problem. Burnout is real, but worse than that in today's world we invariably continue to push through, and chronic diseases occur.

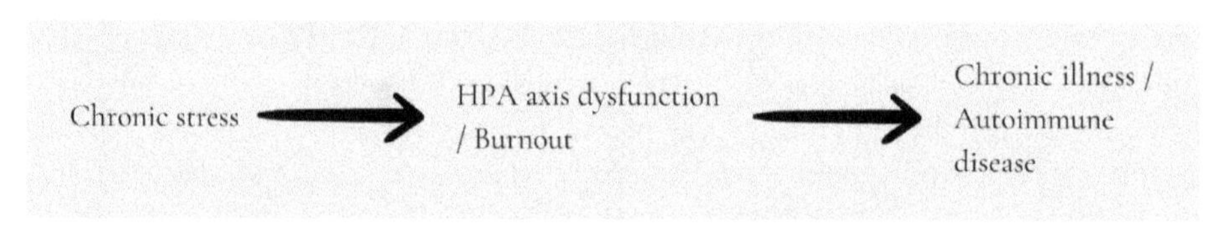

We were not designed to live in perpetual stress. Stress hormones downregulate genes to create disease, and burnout is just the first stage of this.

Second is the fact that while I love supporting people on their healing journey with autoimmune disease, I could also see in so many of the people I worked with that if I had got to them sooner, if we could have jumped in earlier, if we had worked in a preventative way, their story and outcome could have been very different. This is also the case for my own health journey (more on this in chapter 5), but it was this that sparked an interest in me to look at how I could raise awareness around the need to optimize your health.

What is cellular energy?

When we are talking about energy we are often referring to cellular energy. Cellular energy is the currency that powers the biochemical processes that are essential in our body for cellular function, metabolism and overall health. The key to this is a molecule called adenosine triphosphate (ATP) which is the primary source of energy in our body, the currency if you like. We will learn more about ATP in chapter 6. Cellular energy production occurs via a process called cellular respiration, which is a series of biochemical reactions taking place in the mitochondria (chapter 6). Through aerobic respiration the mitochondria generate ATP. This cellular energy gives our body the power to carry out the wealth of processes it must do every second of every day. To produce this energy your mitochondria need nutrients from your food and oxygen. Your nutrition plays a vital role here.

What is immune resilience?

Immunity is your body's ability to effectively respond to and defend against various pathogens, toxins and other challenges – people often talk about boosting their immune system. Immune resilience focuses more on balance. Immune resilience involves a well-coordinated and balanced immune response, where the immune system can distinguish between harmful invaders and the body's own cells. It also includes the ability to modulate the immune response to prevent excessive inflammation, or autoimmune reactions. We don't want a boosted immune system, we want balance – this is immune resilience (more on this in chapter 7).

How many times have you heard someone say: 'The kids keep making me sick' or 'It's just an age thing'?

These are the two main excuses we always use, right? The fact is, your children have developing immune systems and are coming into contact with viruses and bacteria they haven't come across before. Getting sick allows their adaptive immune system (more in chapter 7) to grow and develop. You have likely come up against these bugs before, so you should have some immune resilience. It's not your children's fault if you haven't worked on building your immune resilience, or you have lived a life that has impacted your immune function.

Excuse One: 'The kids keep making me sick'

Recently I was working with patient C on optimizing his health. We worked on his immune function, gut health and adrenals. A few months later I'm working with his wife, patient E. Why? Because in her words the children now think their dad is superhuman. While the whole house has gone down with repeated colds and bugs from school, the dad, patient C, has been fighting fit. Now the wife, patient E, admitted that she was dubious about working on gut health to improve her overall health, energy and immune function – she wasn't convinced it would work. But the proof is in the pudding, and now she wants to be as the children say... superhuman too! (Or at least in optimal health.)

This couple also had yearly BUPA health checks and a year on from working with me their health review was much improved. Here is the message from patient E:

'Finished our medical at BUPA and the doctor was amazed at our results and we've mentioned that we have you assist us and made some changes... I said hire my Nutritionist. She's awesome.'

This was all working on balancing the body and building immune resilience.

Excuse two: 'It's just an age thing'

Often, it's not just an age thing. It isn't just inevitable; it's not just a result that you have to accept and allow. Let's look at Alzheimer's disease. Many often put this down to age but it's actually not; it's not a given that as you age you will develop Alzheimer's – if it was, we would all get it. It's also not a gene thing – less than 1% of cases are due to genetics. The good news is that this means there are steps you can take to benefit your brain health (chapter 8) and stave off diseases like Alzheimer's. Simple changes like diet and lifestyle help keep you optimally healthy.

Why is optimal health important?

In the world we live in today, too many of us live on the treadmill of modern life. Everything these days is fast paced; we have access to everything at any time. Your grandparents would finish work at 5.30pm, go home and have nothing to do with work until 9am the next day. They didn't have phones, email, social media or anything else to interrupt their downtime. Downtime was downtime. Now, can you ever really say you switch off?

Since the Covid-19 pandemic we have all become more acutely aware of this. We were forced to step off the treadmill, to slow down. Quite honestly I think most people quite liked that aspect of it. The benefits of a slower paced life became clear.

Many of us have realized that we can't keep pushing through the way we were doing; we've realized that was actually making us ill, or at least sub-optimal in our performance. We suddenly realized that fatigue and brain fog that we often class as normal, or a sign of ageing, don't have to be our norm. Our performance can remain optimal for longer. People started to look to alternatives as we have all become more interested in health – our own and that of our family. Preventative medicine came to the forefront and the impact of stress became clear.

PEARL: Let's look at some of the figures on stress, our fast paced lives and the impact of not putting our health first:

$1 trillion a year is the cost to the global economy on lost productivity due to stress and anxiety alone (much of that cost falling on employers).

That cost is £28 billion a year in the UK.

There are 23.3 million sick days a year in the UK due to stress and burnout.

Optimizing your health not only protects your future, but it allows you to reach your full potential in your present.

The health journey

With so much information out there these days the health journey can be overwhelming and confusing. It's difficult to know what information you can trust. Looking to genuine sources is really important, as there's a lot out there these days from influencers who actually have no training in what they promote. There's also a lot of promotion and advertising around general health advice that isn't suitable for everyone. If you want information about nutrition look to a nutritional therapist or nutritionist – make sure it's one that is registered. Anyone can call themselves a nutritionist, even if they only did a three-week course. Do your homework, learn about your body and learn about food.

It's also important to remember that your health journey isn't a straight line. It's an ongoing process that will change and fluctuate over your lifetime. When you are healing there will be setbacks, your body fluctuates, outside factors (such as times of stress) will come into play – don't be upset or frustrated about this. Making changes that last you a lifetime are the key to success and over time you build that resilience to withstand more, suffer less setbacks and keep yourself healthier. This isn't a quick fix, it truly is a journey, but it's a very exciting one when you start to see the results.

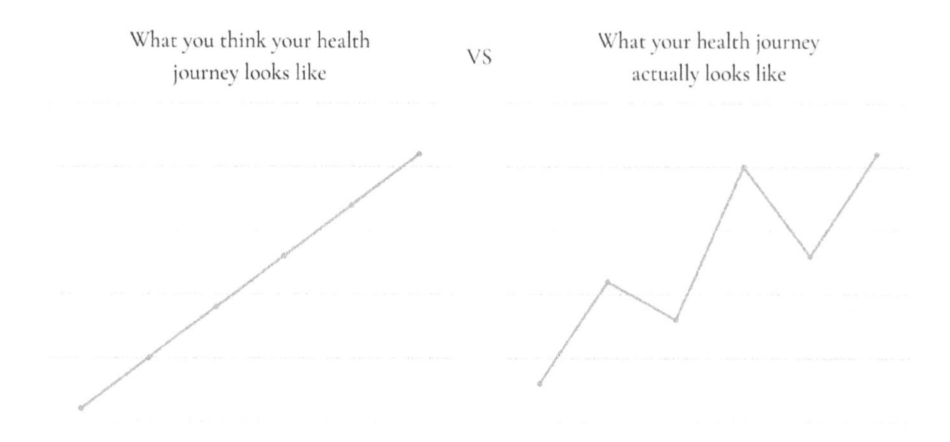

How to use this book

I've specifically designed this book so that you can choose to either read the whole book and jump into everything, or you can pick out sections and decide to focus on one area at a time.

Part 1: The foundations of optimal health – introduces you to the key concepts we discuss throughout this book.

Part 2: The pillars of optimal health – takes you through each of the eight pillars (mitochondrial, immune, brain, adrenal, thyroid, gut, hormone, cardiometabolic) that help you find balance in your body, giving you the science, knowledge and understanding you need to know why you should work on this part of your health. I find with everyone I work with that if they understand why they are making a change they are more likely to make it and stick with it. At the end of each chapter in Part 2, you will find a table that tells you which of the nutrients impact that part of your health, you will find more information on the nutrients in Part 3 when we start building out your plan.

Part 3: The MitoImmune Way – your path to optimal health – is your actionable plan including a 14-step nutrition plan that supports all the pillars of your health. Then we deep dive further, looking at specific nutrients to support each pillar; you can refer back to the checklist at the end of each chapter in Part 2 to see which nutrients you need to focus on for each pillar. We also look at the 4 Lifestyle S's, the quality of our food, functional medicine testing and what you can do next to further your health journey.

Throughout the text you will find PEARLS (you have already seen a few) – these are facts or statistics of particular significance or importance. You will also find case study boxes taken from my work with people just like you.

To support your journey through this book and the implementing of the plan, I've created an online book resources centre for you, with access to lots of tools, recipes and extra information. You can access the online book resources at www.nicolegoodehealth.com/optimal-you-book-resources. Password to access is OptimalYouBook.

If you want to assess which pillars of your health you should personally work on, first take my free MitoImmune Health Assessment online (www.nicolegoodehealth.com/mitoimmune-health-assessment). You will receive a personalized assessment of your health pillars and the areas you score the lowest in are the chapters in Part 2 that you should start with. I advise you to take this assessment anyway, even if you want to read the whole book and work on it all, as it allows you to grasp your personalized health journey and plan. As you will see in chapter 4, personalized medicine is key to optimal health.

So you have a choice, you can either:

- take the assessment and work on your lowest scoring pillars first, dipping into those chapters in Part 2, implementing the 14-step nutrition plan in Part 3 and then looking at the specific nutrients for that pillar of health, or
- take the assessment, read all the chapters in Part 2 and start working on the whole MitoImmune Programme in Part 3. As you work through the book you will see how interconnected all the areas of the body are anyway.

Remember, don't get overwhelmed. *Small steps = Big change.*

Before you read on take my FREE MitoImmune Health Assessment at this link: **www.nicolegoodehealth.com/mitoimmune-health-assessment.**

To access all the book resources head to: www.nicolegoodehealth.com/optimal-you-book-resources.

Password to access is OptimalYouBook

2

Functional medicine

Functional medicine is a science-based, personalized, preventative approach looking at the fundamental underlying causes of clinical symptoms. It focuses on the root cause of the disease, while traditional western medicine places its focus on treating the symptoms. Functional medicine practitioners want to find out what is causing an individual to have those symptoms by looking at your health history to identify triggers – this can help us find out what is creating the problem in the first place. We can then use a set of tools in a strategic way to unravel the complexities of chronic symptoms, or illnesses. It is a simple case of cause and effect: manage the cause and you will see an effect on the symptoms.

Functional vs conventional approach

Now, before I start on this comparison I want to make it clear that both conventional and functional medicine have their roles to play. This isn't a case of choose one and you can't have the other. Conventional medicine has its place and I work alongside many brilliant conventional medicine doctors, but functional medicine has a role to play as well, especially when it comes to chronic health conditions, or prevention and maintaining optimal health. The review of conventional medicine here is also looking at the system, not an individual doctor, and of course finding a truly brilliant doctor can make all the difference. This comparison is to help you understand where you may find benefit from choosing functional medicine as many don't understand the concept. So let's look at the differences and where each approach succeeds.

Emergency care vs chronic care

The conventional medicine system was designed as an acute care system and it works amazingly well for emergency care, trauma and life-threatening care. If you have a heart attack or a road accident you don't want functional medicine, you want conventional medicine, a hospital and likely drug therapy or surgery.

The functional medicine system is designed to support chronic health conditions and support optimizing health as a way to prevent and heal chronic health conditions.

Symptoms vs cause

Conventional medicine focuses on symptoms and diagnoses. What symptoms does a person have, what disease does that mean they have and what treatment do we give that disease to suppress symptoms?

Functional medicine focuses on cause. What is the root cause of the symptoms for that individual? What can we do to restore balance for those root causes so the individual doesn't experience those symptoms? It also appreciates that there can commonly be more than one root cause impacting the person's health.

System-based vs whole body approach

Conventional medicine takes a system-based approach: one doctor for each body part, multiple departments may be required – unless you find a consultant general physician, who are rare these days, but my family have at times been under an incredible one and if you do find one they are amazing to have on your team.

Functional medicine takes a whole body holistic approach, bringing your case under one roof and assessing the interconnections around the body and how they are impacting everything else.

Disease centred vs patient centred

Conventional medicine has treatment plans based on the disease it is treating.

Functional medicine looks at why that individual is having those symptoms. The same disease may have many different root causes and the root cause may be different in different people with the same diagnosis, therefore they will require different care. As a patient once said to me '***you are like a health detective***'.

What vs why

Conventional medicine asks WHAT a person has, functional medicine asks WHY the person has it. We look to take a full health history from preconception, when you were in your mother's womb, to the age you are now.

What is health

Conventional medicine views health as the absence of disease. Functional medicine sees health as a state of optimal wellness.

Fast vs dedication

Conventional medicine can provide quick results, which is especially beneficial in those acute care situations. Surgery or drugs can act fast to get results in life-saving situations. Functional medicine has a slower impact that requires patient dedication. A patient must be a proactive participant in their care and make the necessary changes in order to see results.

Reactive vs preventative

Conventional medicine supports you once you reach a state of disease – at this point you have irreversible damage. Functional medicine promotes preventative care, jumping in as early as possible to avoid irreversible damage where possible.

In summary

So functional medicine is a patient centred, science-led, integrative, holistic, root cause focused approach for chronic care.

PEARL:

50% of adults have at least one chronic health condition (60% in the US).

25% of adults have two or more chronic health conditions (40% in the US).

86% of all healthcare costs are on chronic health conditions.

Functional medicine helps with healthcare costs, burdens on healthcare systems and long-term outcomes for chronic patients as it works to get to the root causes of the individual, allowing them to restore balance (more on this in chapter 3).

Over time the costs of things usually come down; look at tech as an example. Tech becomes cheaper – to run an iPad in 1940 would have cost a lot more than it does today. Technology has moved on. Solar panels are now cheaper to make and run; you can even have them on your own home. In healthcare, costs to sequence the human genome have absolutely plummeted since 2001 from $100 million to around $1,000. Yet the cost of looking after one person's health has absolutely skyrocketed.

PEARL: Poor health costs in 2022 reached $9 trillion globally.

When we as a global society are advancing, how can we be failing so badly at healthcare? Our conventional healthcare systems are not focusing enough on root cause. We will look more at root cause medicine in the next chapter, but the important thing here is that instead of suppressing symptoms (potentially with drugs that will give more side effects and need more suppression of symptoms) there are many health issues where we need to actually get to the root cause of the problem. ***We need to heal instead of suppress***.

Cellular health, telomeres and epigenetics

To truly understand how to support our health we must go down to a cellular level. You need to understand three things here: cellular health, telomeres and epigenetics.

The health of individual cells collectively determines the vitality of tissues, organs and systems within the body. Cellular health involves a delicate balance of energy production, waste elimination and efficient communication between cells.

You know the little plastic things on the end of a shoelace that stop the lace from fraying? Well imagine the shoelace is your chromosome of your DNA and the plastic cap is the telomere. They stop the DNA from fraying, which means that you get accurate cellular replication. However when cellular division happens these telomeres shorten, effectively accelerating cellular ageing. This impacts our overall health and leaves us more at risk of developing diseases.

Now while genetics are one part of the picture, we have to consider epigenetics. Imagine that your gene is a light switch, it can be turned on or off. This action of turning a gene on or off is epigenetics; it is how the gene acts, how the gene is expressed. So epigenetics determines whether a gene is activated or silenced, on or off, just like the light switch.

You see it isn't just our genes that matter. Our genes are the fixed code, they can't be changed, but this is part of the picture. It is how the genes are expressed that matters, the epigenetics, which can steer our health story. So you may always have the genes for a certain disease, but you could go through your whole life with those genes never switching on.

This is where it gets really exciting. Genes are fixed. Gene expression, also called epigenetics, is flexible. It can be changed. What helps modulate the gene expression? Our environment, aka our nutrition, sleep, stress, movement, infections, toxins, etc. The beauty of this is these are all things we can work on to improve our health, to modulate our gene expression, to flick the light switch.

Prevention is better than a cure

> The doctor of the future will give no medication, but will interest his patients in the care of the human frame, diet and in the cause and prevention of disease.
>
> ~ Attributed to Thomas A. Edison

Of course we still need medication sometimes, but preventative medicine was the future, now it is here, you can work on your health in a preventative way to live in optimal health, so that you can achieve all the goals and success you desire.

'Prevention is better than a cure' is usually attributed to Dutch philosopher Desiderius Erasmus in around 1500. It's amazing how much they knew back then that we seem to have lost sight of now.

If you aren't working on keeping yourself in optimal health, then you are heading towards chronic illness. Our health isn't static, it's constantly fluctuating. At one end of the scale, we have optimal health and the other chronic illness, but it's bidirectional, you can move either way. If you get to chronic illness you have to hope there is a cure, but in reality chronic illnesses have no cure, hence they are chronic. You then are forced to devout a lot of time to your illness. Doesn't it make more sense to give a little time to keeping yourself living in optimal health? Give some time now to keeping yourself on top of your game, rather than being forced to take more time later to deal with illness. Supporting immune health isn't just about helping people with chronic illnesses, it's about helping prevent you from ever getting there and if you already have a chronic illness, especially something like an autoimmune disease, it's about building tolerance to help you move towards remission and not trigger further autoimmunity.

Think of food as messages to your cells, it can turn things on or off. It is so much more than just satiating hunger, or giving you energy.

Your timeline

In functional medicine you will often hear us talk about the functional medicine timeline. This is a method we use to assess your health history. The first consultation I have with people is a functional health assessment, and part of this is the building of the patient's timeline to understand their case and look for patterns. Here we are starting to look for root causes (see chapter 3).

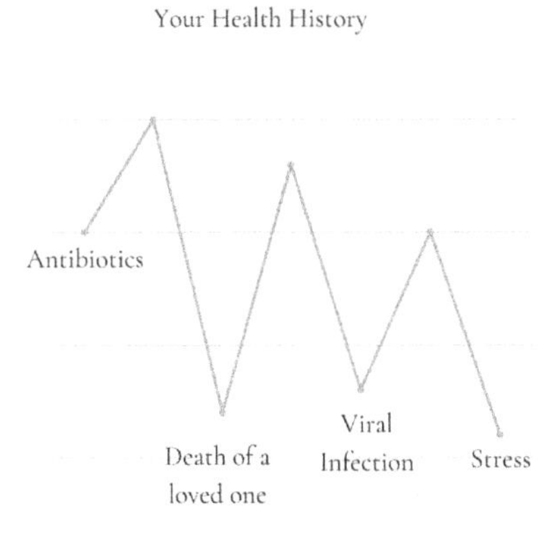

Why is it important to look at your health history? As we saw on page 12, your health journey is an uphill trend if you work to optimize your health, but it's not a straight line; your health history is the same. From your health history we can look at times where your health declined and try to work out what was going on at that time.

If we look at when symptoms started and what was going on for you at that time in your life, or potentially in the months or years prior to the symptoms starting, we can start to see patterns.

I want you to have a go at this yourself. In the following diagram you will see a sample of a filled in timeline; this will look different for everyone as it is your unique story. In the book resources section on my website (www.nicolegoodehealth.com/optimal-you-book-resources) you will find a blank copy of the timeline. Print one out and write on it, or use a tablet to draw on it. Spend some time thinking about your case, your symptoms, your life and just start getting it down on the timeline. This can also be a great tool to take with you when seeing doctors as it can help to organize your thoughts and your story and allow you to communicate effectively about your case.

So how do you fill it in?

1. You will see along the middle a big arrow – in the head of the arrow write your current age.
2. Along the arrow you will write the age you were at the time of symptoms starting, or life events happening.
3. Symptoms will sit below the arrow – draw an arrow down from the age you were at the time and add your symptoms.
4. Life events will sit above the arrow – draw an arrow up at the age you were and add any life events you can think of. Remember this should be at any age not just the ages that correspond with symptoms as there can be gaps between triggers and symptoms starting.
5. Under the head of the arrow write your current symptoms and above the head of the arrow write any current life factors you feel are impacting your health.

Here's an example of a filled in timeline – yours may have a lot more on it, this is just an example.

Head to the book resources section now and print off the prettier version of our functional timeline and tell your story. Remember above the line draw any major live events, stressors or potential triggers. Below the line write symptoms at the time they started, or diagnoses as they were given. Write your age or the year along the middle. Build your health picture.

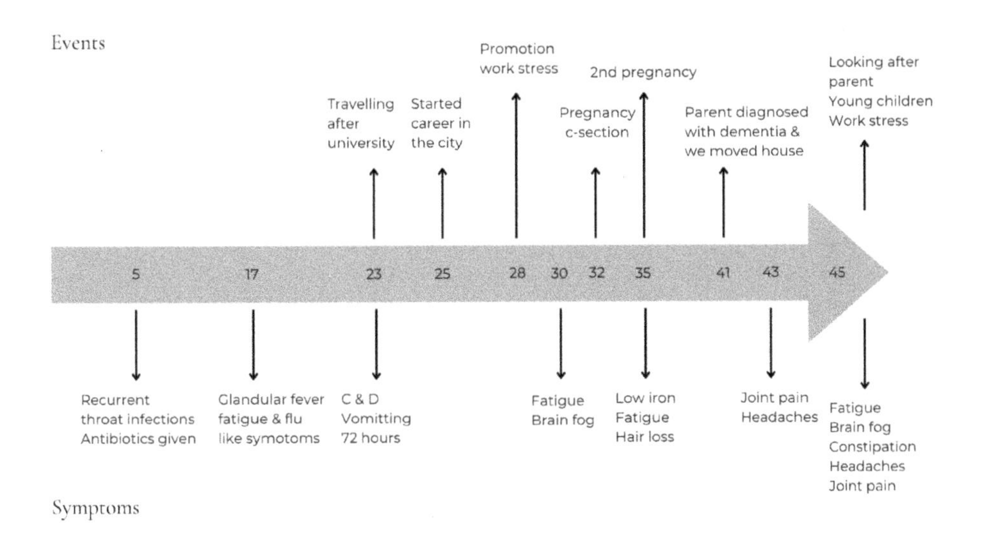

Now you've got your timeline we can use it to look for patterns. So if we take the example given, let's see if we can pull a couple of examples from this timeline. One pattern we see is recurrent antibiotic use as a child and severe food poisoning in a foreign country at 23. So gut health will be playing into this picture. There could be an imbalance in the microbiome (more on this in chapter 11) or some low level infection, or parasite, left over from the travel bug (more in chapter 11). We can also see a connection between the promotion as a potential trigger to the fatigue and brain fog starting at age 30, so we are looking at the adrenals (chapter 9) and the mitochondria (chapter 6). There were then repeated stressors, work, promotion, two pregnancies and a sick parent, after which the joint pain and headaches started. Here we can start thinking of adrenals and their impact on the mitochondria, the brain and the immune function, for example. In the coming chapters you will learn more about these links around the body; as you do, keep referring back to your timeline and thinking which areas could be impacted for you. This is a simplistic timeline and a simple look at the connections, but hopefully it shows you how to start reading your story. If you work with a functional medicine practitioner this is something that they will do with you and take even further.

3

Addressing root causes

We describe root cause medicine to people by looking at the functional medicine tree: at the top of the tree, where the leaves are, we find the different specialist areas of medicine; moving down the tree to the roots we find the causes that impact your health.

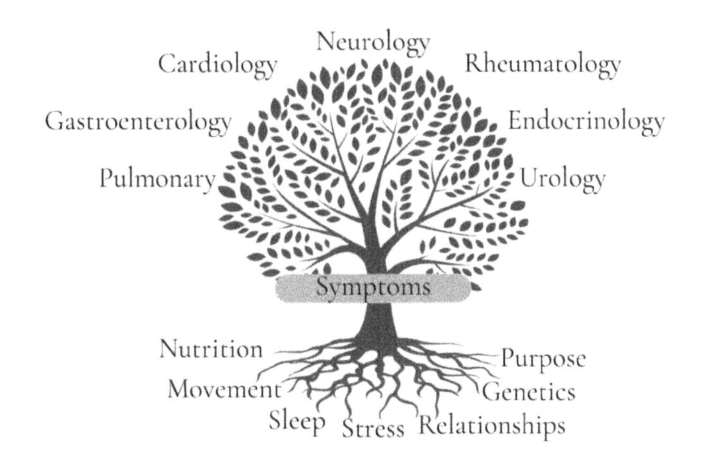

Functional Medicine Tree

Beyond symptom management to the root cause

In functional medicine we look beyond symptom management into the underlying factors that contribute to the manifestation of disease or sub-optimal health. To identify and address root causes is a key component. It is a systems biology based approach to root cause medicine. Identifying and addressing root causes is the secret to optimal health.

Let's look at it through the example of something as simple as a headache. The symptom management approach goes like this:

What have you actually done here? Nothing really, all you have done is take something to mask the pain. There is no reason why that headache shouldn't just come back tomorrow (or in four hours). Now what if instead we use the question 'why?' instead of 'what?', as we looked at in chapter 2?

One condition; many causes

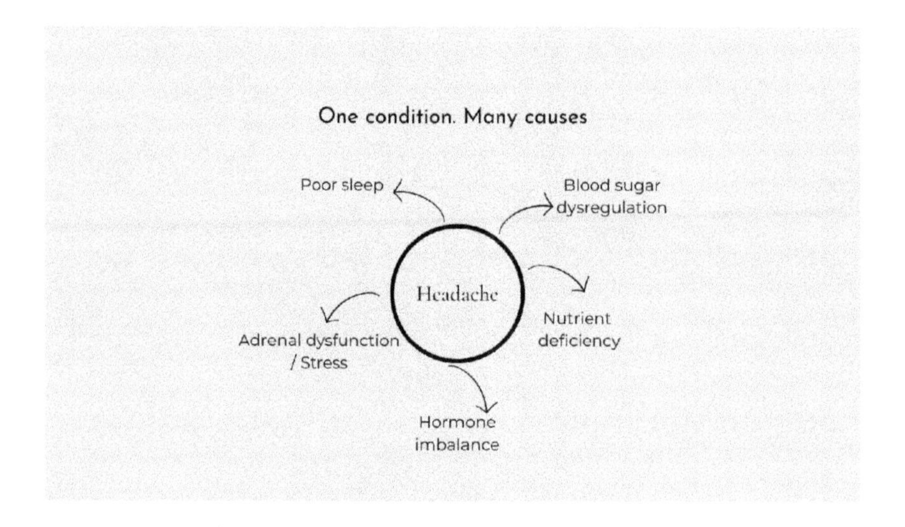

If we look to the root cause we find out why; then maybe we can make it so that the headache goes and never actually comes back. Let's say someone has low magnesium, levels and that is why they have a headache; if you balance the magnesium, the headache and the future headaches are gone. More than that you will fix a lot of other problems in the body caused by the low magnesium; this is root cause medicine. A symptom is just your body's way of telling you something is wrong – don't just mask it, listen to it. Why plaster over something when you could actually fix it?

We can also look at this another way: one cause, many conditions. If a person has inflammation in the body it can cause many conditions.

One cause; many conditions

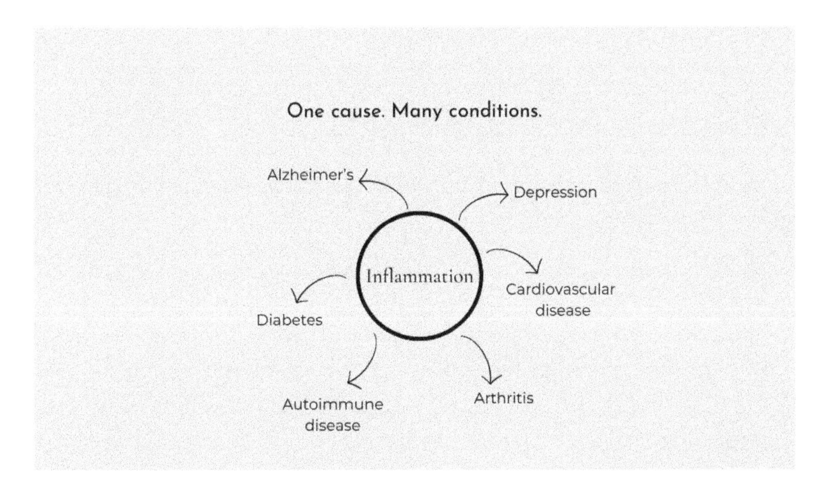

A diagnosis is just a label we give to a batch of symptoms. It isn't the cause, it is the outcome. It doesn't tell us why you have it, it simply tells us what it is. The what isn't as helpful as the why. I could have five people stood in front of me, all of them with the same diagnosis. Let's say they all have multiple sclerosis (MS). That is the what. Great, good to know, it's nice for people to be able to understand what they have, and it may be required for treatment for the what. That's only one part of the picture. So these five people now all have the same diagnosis and are offered the same treatment.

Now let's ask why?

Patient 1:	Underlying low viral load of reactivated Epstein Barr Virus (EBV) and Human Herpesvirus 6 (HHV-6).
Patient 2:	Childhood trauma and low level chronic stress from high powered job.
Patient 3:	A bacterial infection of mycoplasma pneumoniae.
Patient 4:	A highly processed, inflammatory diet is consumed.
Patient 5:	Gut dysbiosis is found with particular presence of clostridium perfrigens.

All of these things are linked to MS – each patient with the same diagnosis has a different root cause. The work I did with each of these patients was very different – each of these five people with the same diagnosis would get a very different plan to follow. This is root cause medicine, which is less focused on the what and more focused on the why.

Now in reality a person is likely to have more than one root cause; this is a simplified view of these cases. Someone may have viral EBV, low chronic stress, poor sleep, low vitamin D and more, but you get the idea.

Removing or balancing all of these things that trigger or exacerbate illnesses helps reduce symptoms and risk, whether it's in someone who already has a diagnosis, or someone wanting to avoid getting one.

Antecedents, triggers and mediators

So let's look at how this works. In functional medicine we talk about ATMs, antecedents, triggers and mediators.

Antecedents (A) are an individual's predisposing factors; here we are looking at things you have inherited like your genetics, your family history, your mother's health preconception and through pregnancy and things you have acquired, such as your birth history, whether you were born via the birth canal or via c-section, your nutrition as a baby and infant, whether you had any illnesses or antibiotics at a young age and what your environment was like. These are all inherited and acquired antecedents, they are your predisposing factors.

Triggers (T) are things which activate a health response: it is a significant event usually lasting a short period (seconds to days), but it has a defined timescale and after such an event your health has been impacted. This could be a trauma, an infection, an injury or a loss of a loved one, for example.

Mediators (M) are something which keep you in a state of sub-optimal health. They are current factors that perpetuate that dysfunction, either in an ongoing or recurring way. This could be ongoing chronic stress from a job or other factor, poor diet, loss of sleep, low levels of movement, an environmental exposure such as to a mould and more.

So whether you have a diagnosis already, or you are searching for optimal health, these play a part. Maybe you have an autoimmune disease in the family so you have the genetics (A), then you had an infection that acted as the trigger for an autoimmune disease (T) and now you have chronic low level stress (M). Or maybe you are well, but just not the optimal version of you. Maybe you were born by c-section so didn't get the good bacteria from travelling down the birth canal (A), then maybe you travelled abroad with work and had a bit of an upset stomach for a few days (T) and now you have a family to support and a high powered job you love, but which is stressful (M).

Either way – diagnosis or searching for the optimal version of you – your ATMs tell your story and are a key part of getting to the root cause. In functional medicine we build a timeline and a matrix for everyone – I do this for everyone I work with to find out their why. The matrix is a way of assessing your systems and how they are interconnected. If you work with a functional medicine practitioner they will likely use this tool.

4

The power of personalized plans

Functional medicine creates a shift in healthcare to emphasize personalized medicine. We've already seen, in chapter 3, the many ways a person can be different and this is where the concept of personalized medicine comes in. Personalized medicine looks at the biochemical makeup, genetic predisposition and lifestyle factors of each individual. Using this information, unique to each person, we create personalized plans rather than using predetermined protocols.

Our health is heavily influenced by our nutrition, lifestyle and environment. In fact more often than not a lot less can be attributed to genetics than you might think. Some research suggests that across all healthcare, 60% of illness, symptoms or diagnoses can be put down to environment, lifestyle and nutrition, compared to just 30% attributed to genetics. The rest is made up of things like access to care. Research into Alzheimer's has shown that in most cases there is no genetic cause; it is purely nutrition, lifestyle and environment. The Alzheimer's Association says there is one gene that can increase your risk (the APOE gene), but that doesn't mean you will get it, and only 1% or less of Alzheimer's cases are from genes that directly cause the disease.

This shows the importance of nutrition, lifestyle and environment.

Functional medicine recognizes that every one of us has our own unique genetics, biochemistry, life history and environmental exposures. This uniqueness is why personalizing medicine is so important. Instead of standardized protocols or treating symptoms in isolation, a personalized approach seeks to uncover your personal story to find your personal root causes, imbalances and dysfunctions to create a plan uniquely tailored to you.

By spending time with a person and creating these personalized plans, we aim to restore balance, vitality and resilience for each individual.

Functional testing

Another way that we personalize care is via science-backed functional testing. Functional medicine testing provides an especially unique insight into your nutritional, biochemical and functional fingerprints. Functional testing is advanced testing not available through conventional routes.

Functional testing often looks at many more markers than conventional testing (learn more about functional testing in chapter 19). Where conventional testing gets to the diagnosis, functional testing gets to the root cause. It also provides much more detail through its many markers that allow us to build personalized plans. Functional testing can also provide you with a 'why'. On paper this can be especially important if you are someone who has been told many times that 'everything is fine', or that 'your test results look normal'. In this way it can also provide hope. Or if you are someone, as many high achievers are, who works well with statistics, numbers and seeing things written down on paper, it can give you a reason in black and white to make a change. Functional testing also allows us to save time and get straight to the root cause, whether this is by finding the root cause, or ruling things out. Functional testing can seem expensive, but in the long run I find that people spend less when they start out with the testing. This health journey isn't a quick fix, it's exactly that, a journey. One that will take time and require change, but will be hugely rewarding.

Why personalized plans matter

WHY PERSONALIZED PLANS MATTER

Superior outcomes
Personalized plans have been shown to give better outcomes for the patient than standard procedures. By tailoring plans we can achieve more targeted and sustainable results.

Empower the patient
Personalized plans have been shown to empower the individual, by involving the person in their healthcare and decision-making they become an active participant providing them with the tools to create long lasting change. Functional medicine personalized plans also promote collaboration between the patient and the practitioner building trust and self sufficiency.

Precision
Personalized plans take the principles of precision medicine tailoring targeted approaches to improving health. Precision medicine contrasts a trial and error approach.

Prevention and early intervention
Personalized medicine plays an important role in preventative medicine and intervening earlier to avoid further damage. Prevention allows for identifying potential risk factors, optimizing health and reducing risk of disease. Early intervention significantly reduces the burden of chronic illness.

Holistic care
Personalized plans take a holistic approach looking at physical, emotional, mental, and environmental factors, recognizing the interconnectedness of the body systems allowing us to...

PEARL: ... view the person as an individual, but the body as a whole.

5

The MitoImmune Method

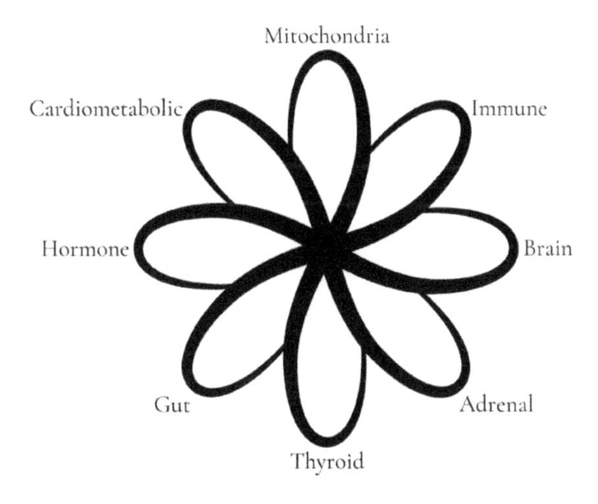

The MitoImmune Method is my signature method that I designed based on my clinical work and training. It is designed to support energy and immunity. The importance of both energy and immunity became apparent to me at a young age, following a viral trigger (Epstein Barr Virus). I had chronic fatigue at a time when I had a lot of stress. It was not stress that was unusual – I was at a very academic school, I had done 11 GCSEs and was studying for five AS Levels and four A Levels. It was standard procedure for most 17 year olds and the stress alone would have been just that, exam stress that dissipated when exams were over – we are built to handle short-term stress and our body does this very well with the age-old fight or flight response, which we will go into in chapter 9. But that stress for me mixed with the viral trigger at just the right time, and my underlying genetics was enough to set me on a path to autoimmune disease. This path was 15 years long before I got a diagnosis. It was one filled with misdiagnoses, medications that didn't work, even a surgery it turned out I didn't need. So let's be honest, the stress didn't really stop. Neither did I. I was off to university to study neuroscience. I wasn't likely to slow myself down easily, that just isn't my personality. I was never pushed by family, but I always pushed myself. Even when I knew the stress had led to burnout, which led to chronic fatigue and the virus, I kept going and pushed through until I really couldn't any more. Even now I still do that, when I know really I should rest. I have to actively watch myself with this all the time – I know all the steps and everything I should do, but it goes against my grain, so I have to work hard at this still. Health is a journey; it's not an overnight win, or a magic pill, it's a lifestyle. Many high achiever's and perfectionists are impacted by health issues at some point, it's a part of our personality. I actually found it even harder once I started to feel better – when you have been forced to rest you don't often want to give in to that again. I had so many days when I was on the sofa unable to function, that now if I get ill I find it hard to have a sofa day; my brain keeps saying 'I'm not going back there'. We are learning all the time; none of us are perfect and I still do things like pushing through when I know I shouldn't. What I have learnt now though is to view these things – like sleep, rest, nutrition, stress management – as ways to help me be able to stay on top of my game. They aren't taking time away from me, ***they are actually allowing me to perform better with the time I have***. Thankfully I have the tools and knowledge to manage my health better now, and because it's in a much better place than it was back then, my body can withstand much more pushing, it has more tolerance, which we will look at in Part 2.

What turned the corner for me was finding nutritional therapy, lifestyle medicine and functional medicine. This was prior to diagnosis, but it didn't matter, these approaches started benefitting me almost immediately. I saw firsthand the power of food and lifestyle as medicine. I had originally intended to stay on at med school after neuroscience to do medicine, but my health had other ideas on this. When I found nutritional and functional medicine I knew this was what I wanted to do, so I went back to studying (the third time now, as I did a law degree in-between – I told you I didn't like to let it stop me) and studied nutritional therapy at ION (The Institute for Optimum Nutrition). This training gave me a three-year science-based nutrition programme, plus (and this was important to me) clinical training. This clinical training part of my programme was what separates nutritionist training from nutritional therapy training. I knew I wanted to open a private practice and that is what nutritional therapists are trained for; we have to do clinical study and work in a training clinic, plus undertake clinical assessments and exams. I loved it from the moment I started. My focus was always on immune function and autoimmunity; that was always going to be where I specialized. So many of these illnesses cause such debilitating fatigue that energy was paramount, and the more I learnt about mitochondrial medicine the more things fell in to place and made sense, not only for my own health, but for how I wanted to support others.

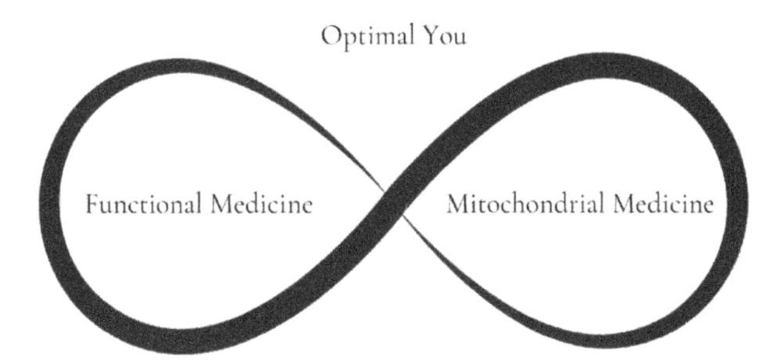

My passion is not just for people diagnosed with autoimmune disease. I've been one of the ones with no diagnosis and endless unexplained symptoms; I've had the fatigue and not known what is wrong; I've been told by the doctors that I'm fine and nothing is wrong. I started working in autoimmune conditions and unexplained symptoms, supporting people to supercharge their energy levels and strengthen their immune resilience (hence the name of this book).

What I learnt the more I worked in clinic was that stress played a huge role as a trigger for these things. Statistics suggest that 70% of people visiting the GP have underlying stress as a root cause. In my clinic I'd say that's more like 90%. Now this isn't to say stress is their only trigger; it's very rare that a person has just one isolated thing going on – the body is interconnected remember, but stress usually features in the history of the patient at the time of, or shortly before, their health began to decline. This is the beauty of doing a full, functional health history. A recent study in *The Lancet* looked at the global impact of diet on health; it had some limitations, but in all it provided a good overview and the conclusion showed that a diet without enough healthy foods and with too many bad foods (predominantly refined grains, sugars, trans fats, junk food) was the number one cause of death and illness, surpassing even smoking.

PEARL: It concluded that in 2017 a poor diet was responsible for 11 million deaths and 255 million years of disability or years of life lost globally.

The more I worked and the more postgraduate functional medicine study I did, the more I realized that stress was leading to burnout, which was leading to chronic illness; I saw it happening every single day. I started to realize that supporting people who had reached disease state was one side of what I wanted to do, but I also wanted to raise awareness for people to jump in earlier, to optimize their health so that they didn't reach disease state. The MitoImmune Method is built with all of these people in mind. Wherever you are on that journey, optimizing and preventing, unexplained symptoms, or diagnosable disease, the Method works for you.

Through years of clinical study and practice I have brought together my knowledge in functional medicine, which includes nutritional and lifestyle medicine, with mitochondrial medicine to identify eight pillars that we can work on to elevate energy, vitality, longevity and performance on a cellular level while balancing immune function and resilience. The method leverages the profound interconnectedness between our cellular powerhouses, the mitochondria and the guardians of our wellbeing, the immune system.

The pillars

We've looked in chapter 1 at energy and immunity, you know my personal connection to it, but why focus on these eight pillars that we are going to dig into in Part 2? The method is based in science, the latest research we have on how to optimize our energy, immunity and longevity. It is also founded in the knowledge that our body is interconnected; we can't work on one singular area and expect to see results. As you work through the pillars in Part 2 of this book you will see all the ways the different pillars are connected and why we need to work on them together.

So what are the pillars? If you have already taken the online MitoImmune Health Assessment you will know, if you haven't you can access this in the online book resources (www.nicolegoodehealth.com/optimal-you-book-resources; remember the password OptimalYouBook) and I recommend you do as it will help you to see which pillars may be most impacted in your own health.

1. Mitochondrial Health fuels your energy and allows you to strive for optimal health and performance.
2. Immune Health builds resilience so your body can avoid infection and stave off chronic illness.
3. Brain Health works on your cognitive performance enhancing your brain power and reducing brain fog, keeping you mentally sharp for longer.
4. Adrenal Health builds resilience to the stress we encounter in the day-to-day world we live in and helps prevent chronic stress from becoming a trigger, giving you back your energy, immunity and performance.
5. Thyroid Health is the second of the endocrine systems we look at and one that is commonly imbalanced as we reach middle age; balancing your thyroid protects you from fatigue.
6. Gut Health, the cornerstone of optimal health and immune function; healthy gut, healthy body – no method to optimize health would be complete without it.
7. Hormone Health – your hormones run the show, and we need balance to function optimally and carry out many functions around the whole body.
8. Cardiometabolic Health is both a preventative part of the method and a known underlying trigger for chronic illnesses; optimizing your health is only complete if your risk factors are reduced.

Now you know what is ahead, let's jump into the pillars and start your journey to the ***Optimal You***.

PART 2

The pillars of optimal health

6

Mitochondrial health

MitoImmune – the 'Mito' in my signature method comes from the word mitochondria to stand for energy, but what exactly are the mitochondria?

PEARL: Your body contains trillions of mitochondria.

The mitochondria are an organelle in the cells of our body within which the biochemical process of energy production and respiration occur. Virtually every single cell in our body has mitochondria in them; think of them as the batteries, the engines that drive all our cells – they are the powerhouses that allow us to produce energy. Energy is required for all the biochemical processes in the body.

The mitochondria do this by converting nutrients from our food and oxygen into cellular energy.

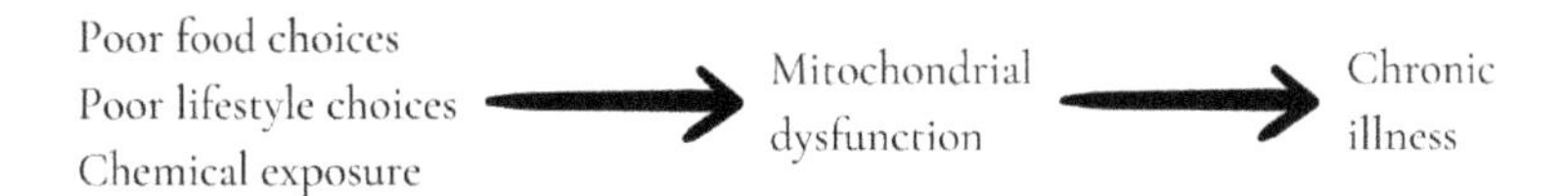

The mitochondria are unique because they have their own DNA – mitochondrial DNA (mtDNA) – which evolved from ancient bacteria that formed a symbiotic relationship with our cells billions of years ago. It is this uniqueness that enables them to generate energy through the process of oxidative phosphorylation. With age, mtDNA experiences damage, resulting in more free radicals and oxidative stress (read more about this on page 38). The more a person protects their mitochondria the better their longevity.

PEARL: There is no vitality, anti-ageing or longevity without optimal mitochondria.

What is energy?

Adenosine triphosphate (ATP) is our primary source of energy and these mitochondria are specialized structures within our cells responsible for producing ATP. ATP that the mitochondria produce is used to maintain your body temperature, for muscles to contract, in hormone synthesis and more.

So how much of this ATP do we need? Well this is where you can really see how important the mitochondria are.

PEARL: We need our own body weight in ATP every day.

Now consider that this is something that fits inside tiny cells in our body. That's a lot of ATP. What makes this even more astounding is the fact that we cannot store ATP in the body. Some vitamins, for example let's take vitamin D, can be stored in the body. If you get enough vitamin D in the summer (many of us don't) you will store it for use over the winter when we don't get enough from the sun. That cannot happen with ATP. We need our body weight daily and we need to make this much every single day. This means our mitochondria must function every second, of every day, of your entire lifetime. Making ATP is of such importance that the mitochondria can account for around 40% of each heart muscle cell and around 25% of each liver cell.

Free radicals and oxidative stress

A by-product of this chemical reaction that occurs to give us energy is the production of free radicals. If you think of the mitochondria as the engine, the free radicals are the fumes that come out of the exhaust. But of course in our bodies we don't want those fumes; the free radicals roam around so they have to be neutralized and mopped up. If they aren't cleaned up they will cause damage that we refer to as oxidative stress; think of this as rust in our engine analogy. So to put this in simple terms, the mitochondria power the engine which produces fumes, which if not eliminated cause rust to the engine and the whole process breaks down.

Free radicals are atoms or molecules that contain unpaired electrons in their outer shell, making them highly reactive. They can be generated from various sources, including normal cellular metabolism, environmental pollutants, radiation and toxins. Free radicals are unstable and seek to stabilize themselves by stealing electrons from other molecules, which can lead to oxidative damage and the formation of new free radicals. This is a vicious and highly reactive cycle. While some free radicals play important roles in cellular signalling and immune defence, excessive free radical production can cause oxidative stress and contribute to cellular damage, inflammation and disease.

Now this cellular damage can be cell death, for example, which can occur in the brain; this has been linked to the onset of neurological conditions like Alzheimer's, Dementia, Parkinson's and MS, as well as being linked to many of the ageing symptoms we have come to expect. But these things don't have to be normalized, and this is where longevity medicine comes into play. We can protect our mitochondria and therefore healthy ageing.

Free radicals is the term you will likely hear used, but you may also hear about reactive oxygen species (ROS) – these are one type, or subset, of free radicals that contain oxygen. Oxygen is used with the nutrients from our food to generate energy (ATP) through oxidative phosphorylation, so when talking about the mitochondria we can say ROS.

Newer research has shown that oxygen and high energy electrons leaking from damaged mitochondria are the primary source of oxidative stress causing cell damage. This is worsened in people who don't have diets which provide enough of the right nutrients to support the mitochondria. This is a positive thing as it means that what we eat really can impact our mitochondrial health.

Antioxidants

We've seen the damage that can occur if we don't look after our mitochondria. The good news is we can do lots to both support our mitochondria and also to counter the damage caused by free radicals.

PEARL: Antioxidants are substances that prevent and slow the damage to our cells caused by free radicals.

You remember how we said that free radicals have an unpaired electron that makes them reactive so they can cause damage, well antioxidants pair up with the unpaired electron, effectively neutralizing the risk of damage from the free radical.

We have both endogenous and exogenous antioxidants. Endogenous are produced in the body and exogenous come from outside the body, i.e. we consume them.

PEARL: Glutathione is your master antioxidant.

Glutathione, much like mitochondria, can be found in almost every cell in the body. However, poor nutrition, stress, lack of sleep, illness, environmental toxins and ageing can all lead to lowered glutathione levels in the body. We will learn more about glutathione in chapter 7.

The balance between ROS and antioxidants is essential for mitochondrial health and cellular function. Coenzyme Q10 (CoQ10), vitamins such as C and E, polyphenols such as green tea, berries and dark chocolate are all powerful antioxidants. Even resveratrol in red wine and red grapes are good antioxidants. There is more on these in Part 3 including how to incorporate them into your diet.

We also have to look at cofactors. For antioxidant enzymes we also need things like selenium and zinc. Cofactors are nutrients or compounds that are required for an enzyme process to occur other than the main substance.

How mitochondria impact our health

The health of your mitochondria is directly linked to the health of your cells, and prevention of illness. Cells in our brain, nerves, heart, gastrointestinal tract and muscles have particularly high levels of mitochondria, because they need a lot of energy to function optimally. This also means they are more at risk of damage and decline in function. Our nutrition and lifestyle choices can support the mitochondria health, or impact the speed of this damage leading to illness.

When mitochondria are functioning optimally, you experience:

- Abundant Energy: Optimal mitochondrial function leads to a higher production of ATP, which translates to increased energy levels, vitality and stamina.
- Efficient Mitochondria: Support better endurance, making it easier to engage in physical activities and exercise.
- Quick Recovery: Mitochondrial health is crucial for post-exercise recovery, as it helps repair damaged tissues and replenish energy stores.
- Resistance to Fatigue: With robust mitochondrial function, you are less likely to experience fatigue and more likely to sustain energy throughout the day.
- Improved Mental Clarity: Sharp cognitive function and enhanced mental clarity are associated with well-functioning mitochondria, as the brain relies heavily on energy to perform its complex tasks.
- Balanced Mood: Mitochondria also play a role in neurotransmitter regulation, which can help stabilize your mood and reduce the risk of mood disorders.
- Efficient Metabolism: Proper mitochondrial function supports a balanced metabolism, making it easier to maintain a healthy weight and regulate blood sugar levels.

The impact around the body

What you will see as you tour through this book is how interlinked the pillars are and why they are all part of my signature method; you can't reach optimal health without working on them all and they are all inextricably linked with one another. If one goes out of balance it will shift another and so on. The interconnectedness of the body is discussed more in chapter 2 when we talked about functional medicine.

So how do the mitochondria impact the rest of the pillars?

Mitochondria and immune function

As with all cells, the mitochondria provide energy to your immune cells, but this isn't where the connection ends. Mitochondria also play an active role in immune signalling, the regulation of inflammatory responses and cell fate decisions.

Immune cells such as T cells, B cells, macrophages and dendritic cells (don't worry we are coming onto the immune system in the next chapter) require huge amounts of energy in order to perform their actions. The energy mitochondria produce is required for immune cell activation, cytokine production and clearance of pathogens.

Mitochondria, as we have seen, are also a major source of ROS. In immune responses ROS act as signalling molecules (as with everything we want balance, we need some ROS); low level ROS can be anti-microbial.

Dysfunction of mitochondrial metabolism can impair immune cell differentiation and function, contributing to immune cell dysfunction and greater susceptibility to infections or autoimmune diseases.

The mtDNA can also activate an innate immune response, triggering inflammatory and anti-viral defences to pathogens, a vital part of your immune function and protection from infections.

Mitochondria–brain connection

The brain is one of the most energy-demanding organs in the body, using about 70% of the ATP. It requires a constant and efficient supply of ATP to carry out functions such as thinking, memory formation and maintaining overall cognitive health. When mitochondria in brain cells (neurons) are compromised, it can lead to various neurological issues, including cognitive decline like Alzheimer's disease and Parkinson's disease, mood disorders such as depression and anxiety, fatigue and brain fog and difficulties with concentration and memory.

Supporting mitochondrial function is a vital step for those looking to support optimal health and longevity, for those where the fast paced lives we lead are taking a toll, or where you need to be on top of your game to be able to achieve all you want to do with your life. For anyone with a condition where fatigue is a symptom, you cannot overlook the mitochondria; similarly if you have an inflammatory condition (hello autoimmune diseases), you need to focus on this pillar of health. Let's take a look at one of my patients who recently was struggling with dips in energy and feeling burnt out.

Patient M came to see me with no diagnoses, she generally saw herself as someone in good health, she was a singer and musician, working with a heavy schedule and also travelling a lot for work from California around the world. Her low energy, brain fog and eating on the road meant, in her words, that she was 'struggling to manage everything that needed doing in day to day life'. So what did we do? She had a comprehensive stool test – we found some microbiome imbalance, yeast overgrowth and imbalance bacteria. We worked on balancing this out; we supported the gut by removing what shouldn't be there and repairing the gut lining as she showed markers consistent with intestinal permeability (more in chapter 11) and repopulated the gut with more of the good stuff. She found that her brain fog started to lift and energy slowly improved, but we kept digging as we hadn't yet reached the root cause. Due to all the travel and staying in many different places, we decided we would run a mycotoxin (mould) test to make sure that there wasn't anything underlying there, and sure enough there was. We found significant mould toxicity; she had Ochratoxin A, a persister mould. Mycotoxins released by mould can damage the mitochondria. So we worked on detoxification, further supported her liver and gut and then started work on her mitochondria and brain, and this was where we started to see that final push to optimal health; she really started to feel rejuvenated as we worked on the mitochondrial support. On her results she said, 'In 6 weeks you've given me my life back, a year ago I didn't think this was possible'. She was back to being able to exercise, work, socialize and all within a relatively short space of time. And this was someone who didn't come to me with a diagnosed illness, this was someone who deemed themselves generally well. But often we accept symptoms as just being tired and busy, or linked to age and we shouldn't, we should be able to live optimally.

Mitochondria and the adrenals

Mitochondria play a crucial role in production of our stress hormones (chapter 9). Cholesterol is taken into the mitochondria and converted to pregnenolone, the precursor to the steroid hormones like cortisol. Stress also depletes the nutrients that the mitochondria require to function.

Mitochondria are also involved in the production of the catecholamines, adrenaline (epinephrine) and noradrenaline (norepinephrine), required for the stress response. Mitochondrial dysfunction can lead to dysfunction in the body's stress response.

Mitochondria–thyroid connection

The thyroid hormone triiodothyronine (T3) is required for production of ATP. The thyroid predominantly makes the inactive form thyroxine (T4) and must convert it to T3 (more on this in chapter 10). For this conversion, mitochondrial enzymes type 1 deiodinase (D1) and type 2 deiodinase (D2) are required.

T3 is also required for mitochondrial biogenesis (more on this on page 45 and 78) and some function. T3 can enhance mitochondrial biogenesis increasing ATP production in the body.

Mitochondria dysfunction is a known root cause and contributor to thyroid conditions including hypothyroidism (when your thyroid is underactive), hyperthyroidism (an overactive thyroid), Hashimoto's and Graves (the autoimmune hypothyroid and hyperthyroid diseases respectively). If you have mitochondrial dysfunction the production of thyroid hormones may be impaired as well as thyroid hormone metabolism and the signalling pathways. The inflammation and oxidative stress caused by mitochondrial dysfunction can also contribute to thyroid conditions.

Working on the mitochondria isn't just for people like patient M (in the previous case study box) who deemed herself generally well, free from diagnoses, but struggling with energy and brain fog. It is equally important for those with immune health conditions. Let's look at patient B. Patient B had Hashimoto's and came to me to work on her thyroid. Similar to patient M, we took a whole body approach – we didn't just focus on the thyroid. We worked on gut, adrenal, immune and thyroid health and started supporting the mitochondria. This patient wasn't converting her T4 to T3 well (chapter 10). When this happens the mitochondria slow down and don't function as well. Her thyroid-stimulating hormone (TSH) results on the blood tests and medication were actually pretty good, but she still wasn't feeling as energized as she had hoped. She was being told by the doctor's that everything was ok and couldn't work out why she didn't feel better. What was actually happening was that those inflammatory processes were still at play from the Hashimoto's and this is why it's common to find mitochondrial damage in Hashimoto's patients, due to the free radicals that result from oxidative stress in the body. As we worked on her thyroid conversion and on her mitochondria, her energy levels rapidly increased. She went from not exercising at all to doing three sessions a week, she was coping better at work, she was a mum and was able to really be present with her kids and more. The change in her was amazing to see.

Mitochondria and gut health

The gastrointestinal (GI) tract has a constant demand for energy as the epithelial cells lining the gut wall have a constant turnover. Optimal mitochondrial function supports this process and supports a healthy gut wall and tight junction integrity reducing the risk of intestinal permeability. As well as the gut wall, mitochondria imbalance can impact the gut microbiome. (We will read more about the gut in chapter 11.)

The increased inflammation we have already discussed also contributes to development of things like inflammatory bowel disease, or irritable bowel disease.

Mitochondrial dysfunction also impacts the gut-brain axis impacting neurotransmitter production.

Mitochondria and their connection to hormones

We've already looked at the adrenal and thyroid hormones, but the mitochondria also impact sex hormones and your metabolism. Energy is required for production of sex hormones such as oestrogen, progesterone and testosterone. Mitochondrial dysfunction can lead to deficiency or excess of sex hormones.

The mitochondria are also involved in hormone signalling; dysfunction can lead to altered responses to hormonal cues leading to hormone imbalances and reproductive disorders.

Mitochondria and cardiometabolic health

Mitochondria are critical in regulating glucose and insulin sensitivity in tissue such as muscle and liver. Impaired cellular energy metabolism can lead to insulin resistance, impaired glucose uptake and dysregulation of blood sugar. You can see elevated insulin levels and changes in appetite regulating hormones like leptin and ghrelin.

Mitochondrial biogenesis

Here is the good news: mitochondrial biogenesis is the process by which new mitochondria are formed within cells. It is a tightly regulated process that enables cells to adapt to changing metabolic demands, optimizes energy production and maintains mitochondrial function and integrity. We can support mitochondrial biogenesis through nutrition and lifestyle.

Nutrients and mitochondria

As mitochondria use nutrients from our food and oxygen in order to make this ATP, they are particularly susceptible to nutrient deficiencies and oxidative stress. Eating certain foods can mean that less free radicals are produced and we therefore have less oxidative stress (rust) in our bodies. So not only is cellular energy production fuelled by nutrient dense, high quality foods, but it can reduce cellular damage as well.

At the end of this chapter you will see a list of nutrients to support mitochondria – you will then find all you need to implement the necessary steps in Part 3.

What I want you to understand here is that the mitochondria use the macronutrients in our food – the proteins, carbohydrates and fats – to create the ATP. There are also certain nutrients that are cofactors in the process, such as B vitamins and CoQ10, and nutrients that help reduce oxidative stress such as antioxidants.

As well as nutrients to consume, we have to look at ones that act as anti-nutrients. These are processed foods with unsaturated fats, or hydrogenated fats, burnt food and cured meats.

Lifestyle and mitochondria

Chronic stress and exposure to environmental stressors can negatively impact mitochondrial function and promote mitochondrial dysfunction. Practicing stress-reducing techniques such as meditation, mindfulness, yoga and deep breathing exercises can help mitigate the harmful effects of stress on mitochondrial health and support mitochondrial biogenesis.

Adequate sleep is essential for mitochondrial health and function. Poor sleep quality, sleep deprivation and disruptions in the circadian rhythm can impair mitochondrial biogenesis and function. Prioritizing healthy sleep habits, such as maintaining a consistent sleep schedule, optimizing the sleep environment and practicing good sleep hygiene can support mitochondrial health and promote overall wellbeing.

Physical activity, particularly aerobic exercise and high-intensity interval training (HIIT), stimulates mitochondrial biogenesis by activating signalling pathways.

Engaging in regular exercise routines can increase mitochondrial content and improve mitochondrial function in skeletal muscle and other tissues.

You can find out more about how to implement these lifestyle changes in chapter 16.

The mitochondria nutrients

Mitochondria

NUTRIENTS

- ALA
- B VITAMINS
- CARNITINE
- CAROTENOIDS
- CoQ10
- GLUTATHIONE
- IRON
- MAGNESIUM
- OMEGA-3 FATTY ACIDS
- PQQ
- SELENIUM
- VITAMIN E
- VITAMIN C

7

Immune health

Why I focus on immune resilience

We often talk about immunity or immune function, but what about immune resilience? This is something we don't hear about as much. Immune resilience is the ability for the immune system to be able to quickly launch and defend against pathogens, respond to stress, age and health conditions and then, and this is the important bit, quickly recover, using inflammation to defend and protect you, but then dampening down the inflammatory response to prevent chronic inflammation.

This is a vitally important measure because we often forget that inflammation is a good thing, we need it. Chronic inflammation is the problem. Building immune resilience, or immune tolerance as it is sometimes called, is key to protecting you against chronic conditions like autoimmune disease and other inflammatory and age-related conditions.

Immune health and inflammation

As soon as we talk about inflammation, especially in the wellness world, it is seen as a negative, but inflammatory processes are essential for your defence mechanisms. The pain, swelling and tissue damage you get when you are injured is part of the healing process. Inflammation is part of the acute response that involves activating immune cells, the inflammatory response and activating repair pathways. With acute inflammation, the inflammation flares up quickly and also recedes quickly as you heal. This is a hugely protective and important part of your immune system.

Chronic inflammation, which is what we are talking about when we say we want to reduce inflammation, is a different thing. This is when the inflammation persists, or becomes dysregulated leading to sustained inflammation over a longer period of time. This can be at a much lower level than acute inflammation, but essentially the inflammation doesn't 'switch off' the way it should. This prolonged activation of the immune response can lead to tissue damage, organ dysfunction and systemic health issues.

Immune vs autoimmune

Autoimmunity is a dysfunction of the immune system. Essentially, we lose immune tolerance and our immune systems become less resilient. People often say autoimmunity is your body making a mistake – this is only true to an extent. Your immune system is actually doing exactly what it was designed to do, it is just over-activated and therefore doing too much, which leads to it not being able to effectively distinguish self-tissue (our body's tissue) from non-self-tissue (foreign tissue). Your immune system is being told there is an invader so it is going out to mop it up. This loss of recognition of self and non-self is immune dysregulation and loss of immune tolerance.

What we need to find out is what is driving this. What is the root cause?

Chronic inflammation will be a part of this, and something will have triggered the chronic inflammation – a virus, bacteria, stress, mould, gut barrier integrity, for example, are some of the possible root causes.

PEARL: There are over 100 autoimmune diseases, and more and more are being classed as autoimmune all the time.[1]

As we saw in chapter 1, autoimmune diseases are one of the biggest causes of chronic health issues today and they are rapidly growing.

PEARL: One in 12 women and one in 24 men have an autoimmune disease.

The innate and adaptive immune systems

Let's look at the immune system in a little more detail.

The innate immune system jumps into action fast, responding to anything that comes into your system. Cells like macrophages, dendritic cells and mast cells act as your security guards – think of them as the bouncer on the door of a nightclub.

[1] Have a diagnosis, but don't know if it's autoimmune? This is a common problem. Doctors fail to tell their patients. Head to https://autoimmune.org/disease-information to search your diagnoses and find out. There might be a lot more on that list than you think. Remember this is only the fully scientifically proven autoimmune conditions, there are a lot more that highly thought to be autoimmune and likely to be added to the list.

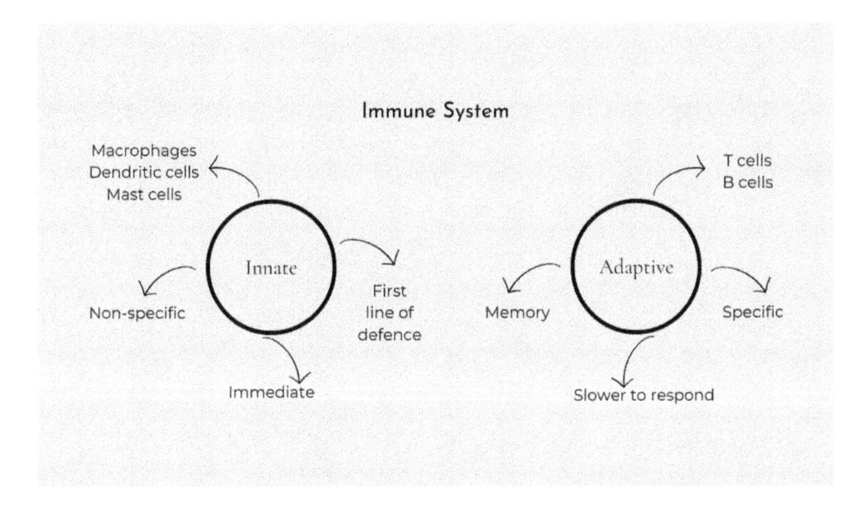

The adaptive immune system is slower to respond and is made up of T cells, B cells and lymphocytes – these are the special forces.

Let's go back to the nightclub. Someone gets in who is causing trouble and we want them out. The security guards, the bouncers (innate immune system), go in and try to remove them. They manage to calm the situation somewhat, putting themselves between the intruder and all the other innocent partygoers, but this invader is tough and so security call the police. The police send in a special forces team (adaptive immune system) to take this person down. They capture, remove and now, super important, they add the invader's profile to the database. Now they will be remembered and known in the future if they try to get in anywhere else.

You see the innate immune system is non-specific. It will deal with any invader up to a point. It's the emergency responder so it can act fast. The adaptive immune system is specific. It has specialized tools and weaponry to target different types of invaders, but it takes time to pull this team together; however, when it gets there it works effectively and, importantly, remembers the invading pathogen, so that if you come into contact with it again, it can act quicker and heal you better.

Why do we need a strong but balanced immune system?

All of this shows us why we need a strong but balanced immune system.

We want strong immunity to keep us well, to respond to pathogens and to prevent us getting ill.

PEARL: However, we want balance, because an immune system in overdrive leads to autoimmunity.

All of these things you read about boosting your immune system are a gimmick. We want balance not boosting. That's why in the title of this book I wouldn't put 'How to supercharge your energy and immunity'; I don't want you to supercharge your immunity. 'How to supercharge your energy and immune resilience', that I'm ok with, because that is saying optimize your immune system to 'switch on' quickly, heal you and then 'switch off' again. Plus how do you know if you are someone who is experiencing a symptom because your immune system is low, or because it's in overdrive? Ageing is also associated with changes in the immune system, known as immunosenescence, which can lead to alterations in immune function, increased susceptibility to infections and chronic low grade inflammation.

The autoimmune spectrum

The medical world tends to look at autoimmunity as you are either healthy or you have an autoimmune disease. In fact, autoimmunity is a spectrum, there are stages, it's not as black and white as you may believe. Often people with autoimmunity are told there is nothing wrong with them. They've had tests where nothing is showing up, so they must be fine, right? Has this ever happened to you?

More often than not, people come to me with symptoms and no diagnosis. Maybe not even something major – a bit more tired than normal, some brain fog, maybe some muscle aches or joint pain. These are all things that could be put down to ageing or a busy lifestyle.

PEARL: Did you know you can have autoimmunity for decades before you would be diagnosed with an illness?

The Autoimmune Spectrum

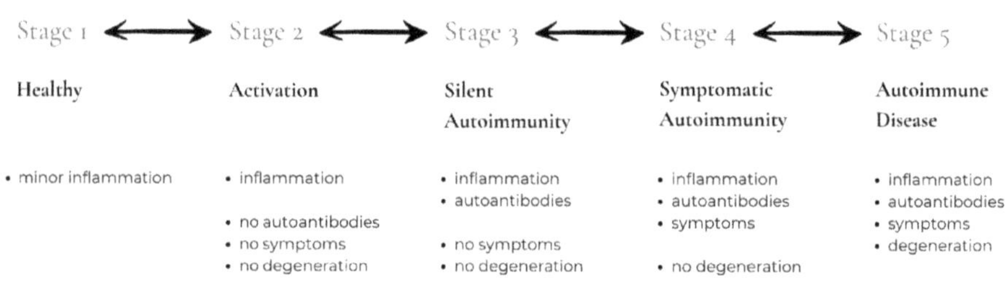

Stage 1 ⟷	Stage 2 ⟷	Stage 3 ⟷	Stage 4 ⟷	Stage 5
Healthy	**Activation**	**Silent Autoimmunity**	**Symptomatic Autoimmunity**	**Autoimmune Disease**
• minor inflammation	• inflammation	• inflammation • autoantibodies	• inflammation • autoantibodies • symptoms	• inflammation • autoantibodies • symptoms • degeneration
	• no autoantibodies • no symptoms • no degeneration	• no symptoms • no degeneration	• no degeneration	

Autoimmunity is a spectrum and importantly you can move both ways along that spectrum – you see those arrows in the diagram, they go both ways.

PEARL: You can go ten years or more in the silent autoimmune phase.

Let's look at patient H. H was a high achiever, always had been, studied hard, worked hard and just getting on track to the career she had been working towards for years. Suddenly she was struggling with fatigue and brain fog. There were a few gastrointestinal (GI) symptoms, so slight that she hadn't even picked up on them until we had our initial consult. I asked her if there were any GI issues and she said 'no' [pause] 'well I guess actually I'm not going to the toilet as often as I used to, and sometimes I get a bit bloated, but no, no real stomach issues.' This is something I hear a lot with my patients… no real issues… what does that mean? It means we have started to accept something as normal, when it isn't and we shouldn't. We all deserve to live optimally well.

H had been to the doctors a few times and had some bloods done, the things you would expect vitamin B12, iron profile, blood sugar, cholesterol, a basic thyroid profile etc., but was told that everything was normal and it was probably stress. Now don't get me wrong, stress is a massive trigger, but it's not the end of the story. H had been offered beta blockers to help with anxiety. Now as I was talking to H, I didn't feel she was overly anxious. I asked her about the conversation she had with the doctor prior to the prescription and she said the doctor asked if she ever felt anxious. She had replied to say that she had a big job interview approaching that was stressful and she had been working hard and long hours, but didn't feel particularly anxious. That was enough for it to be put down to anxiety.

So we assessed her symptoms, took a full functional health assessment, reviewed her family history and found niggling symptoms all around the body. We also discovered quite a family history of autoimmunity, not in her parents, but dotted around both sides of the family. Nothing really serious, but things like Raynaud's where your fingers go white, vitiligo and psoriasis. When we ran an autoantibody profile on her we found that H had autoantibodies flagging up to a few parts of the body. So while she didn't fit into a box for diagnosis, she had inflammatory markers raised, autoantibodies showing and some minor symptoms. She had symptomatic autoimmunity.

Now my question to you is this: would you rather wait until stage 5, when you have an autoimmune disease and therefore damage to a tissue or organ and then work on pushing yourself back down the spectrum towards remission, or would you rather jump in at any of the earlier stages and aim to reach optimal health so that you don't reach stage 5 and don't get damage to a tissue or organ?

It's an obvious answer, right? Preventative medicine is key. But don't lose hope if you already have a diagnosis of an autoimmune disease. Remember that arrow still goes both ways and while you will always have that condition (at least as far as medical science knows at this point) you can push symptoms back towards remission and prevent further damage. Remember that earlier pearl of wisdom in chapter 1: once you have one diagnosis the average is you get five; this is usually because people are not working on pushing back the autoimmunity. Medicines for many autoimmune conditions only work on one part, or to plaster over symptoms; they invariably do not work on the underlying autoimmunity.

The immune connections with the other pillars

Mitochondria

Immune cells rely on mitochondrial energy production as seen in chapter 6. Conversely, immune activation can influence mitochondrial dynamics, biogenesis and function, impacting cellular metabolism and energy production.

Brain

The brain-immune axis is bidirectional. Immune cells, cytokines (proteins used for cell signalling) and inflammatory mediators can cross the blood-brain barrier and influence brain function, neuroinflammation and neural signalling. Immune activation can also impact neurogenesis, plasticity and cognitive function (chapter 8).

Adrenals

The immune system and the adrenal glands communicate bidirectionally through the hypothalamic-pituitary-adrenal (HPA) axis. Immune activation can stimulate the release of stress hormones, while chronic stress and dysregulated cortisol levels can impair immune function and contribute to immune dysregulation and inflammation. Chronic inflammation may also impact adrenal function and cortisol production.

Thyroid

Immune dysregulation is closely linked to thyroid health and autoimmune thyroid diseases such as Hashimoto's thyroiditis and Graves' disease. More on this in chapter 10.

Gut

Over 70% of your immune system is in the gut. The gut-associated lymphoid tissue (GALT) is a key component of the mucosal immune system, containing a large population of immune cells that surveil the gut for pathogens and maintain immune tolerance to commensal microbes and dietary antigens. Alterations in gut microbiota composition, known as dysbiosis, can impact immune function and contribute to systemic inflammation and immune-related diseases.

Hormones

The immune system influences hormone production, signalling and metabolism through various mechanisms. Immune cells and inflammatory mediators can interact with endocrine organs, such as the pituitary gland, adrenal glands, thyroid gland and gonads, affecting hormone synthesis and secretion. Immune dysregulation and chronic inflammation can disrupt hormone balance.

Cardiometabolic

Chronic inflammation and immune dysregulation are major contributors to cardiometabolic diseases such as atherosclerosis, hypertension, insulin resistance and metabolic syndrome. Inflammatory cytokines and immune cells play key roles in the development and progression of vascular inflammation, endothelial dysfunction, insulin resistance and lipid metabolism abnormalities. Immune activation can promote oxidative stress, inflammation and tissue damage in the cardiovascular system, leading to endothelial dysfunction, plaque formation and cardiovascular events.

How nutrition and lifestyle impact immune health

The key with nutrition and lifestyle factors to support immune health is to provide essential nutrients for immune function, balance the immune system and reduce inflammation.

Micronutrients such as vitamins and minerals are essential for immune function; they are cofactors in many of the cellular processes. Vitamin C, vitamin D, vitamin E, zinc, selenium and iron are particularly important for supporting immune cell function, antioxidant defence and inflammation regulation.

As we saw in chapter 6, antioxidants play a vital role in protecting immune cells from oxidative stress and maintaining immune balance. We have abundant amounts of glutathione in the respiratory system (nose, mouth, throat, lungs) – one of our first lines of defence as particles enter the body. To put this into perspective, there is 140 times more glutathione in the fluid lining the lungs than there is in the blood. Glutathione also plays a role in liver detoxification by binding to heavy metals, chemicals and toxins. Glutathione is also necessary to make and maintain white blood cells, T helper cells and Th1 and Th2 cells – all cells used in our immune response. Glutathione also helps by recycling other antioxidants.

Omega-3 fatty acids have anti-inflammatory effects in the body and can modulate immune cell function.

Keeping hydrated is essential for maintaining immune health, as water supports various physiological processes, including lymphatic circulation, toxin elimination and mucous membrane hydration.

Lifestyle changes also play a big role in immune function. Regular physical activity has immune-modulating effects, promoting circulation, reducing inflammation and enhancing immune cell function. Chronic stress can suppress immune function and increase susceptibility to infections and inflammatory diseases. Quality sleep is essential for immune function, as sleep supports immune cell activation, cytokine production and antibody response.

Nutrition and autoimmunity

Gluten

Should everyone be gluten free? I get asked this a lot and the answer is not every one needs to be, however gluten can increase the risk of intestinal permeability (chapter 11). If you have an autoimmune condition, gluten wants to go. So The MitoImmune Nutrition Plan in chapter 14 is not a gluten-free plan; it is immune supporting and anti-inflammatory but not gluten free. You can, however, easily adapt it to make it gluten free and I advise that you do give this a go. If you are struggling going gluten free head to the online book resources on the website where you can access a gluten and dairy free four-week meal plan.

Dairy

Dairy products can elicit immune responses through a process called molecular mimicry in individuals with autoimmunity, leading to inflammation and autoimmune flares in susceptible individuals. You can read more about this in the thyroid health chapter (chapter 10).

Lectins

Lectins – plant proteins found in various foods like grains, legumes and nightshade vegetables – have been implicated in autoimmune conditions due to their ability to bind to cells and potentially trigger inflammatory responses in susceptible individuals. While these dietary components may provoke autoimmune responses in some individuals, personalized approaches to diet and lifestyle modifications are essential in managing autoimmune conditions and promoting overall health and wellbeing. An elimination diet, or autoantibody food testing, can be used to see your response (chapter 10).

Nightshades

Nightshades are a group of plants belonging to the Solanaceae family, which includes commonly consumed vegetables like tomatoes, potatoes, peppers (bell peppers, chilli peppers) and aubergine. Some people with autoimmune conditions may consider eliminating nightshades from their diet due to concerns about the potential impact on inflammation and immune responses. However, it's essential to recognize that the relationship between nightshades and autoimmune conditions is complex, and individual responses can vary. We shouldn't just remove nightshades from the diet as they contain a lot of goodness, nutrients and add diversity to the diet. If after removing gluten and dairy and working on diversity of the diet we don't see improvement we can run an elimination programme to test nightshades; this has to be done in a very specific way in order to get true results (chapter 11), or look to antibody testing and cross reactivity, you need to work with a functional medicine practitioner for this.

Sodium (salt)

Salt activates a pathway which turns on Th17 cells and down regulates T regulatory cells; these are cells of the immune system. The T17 pathway is inflammatory and the T regulatory cells are what maintain balance in the immune system so it is vital these are not impaired in any immune condition. Limiting salt is important. If you do eat a salty meal occasionally you should have potassium with it, or take a potassium supplement.

The immune nutrients

Immune

NUTRIENTS

- ALA
- B VITAMINS
- CAROTENOIDS
- FLAVANOIDS
- GLUTATHIONE
- GLUTAMINE
- IRON
- OMEGA-3 FATTY ACIDS
- POLYPHENOLS
- PROBIOTICS
- PQQ
- SELENIUM
- VITAMIN A
- VITAMIN C
- VITAMIN D
- VITAMIN E
- ZINC

8

Brain health

When I studied neuroscience at university I will never forget the first day in the first lecture. It was the welcome lecture, and the tutor introduced himself, welcomed us and then said 'in the next three years we will teach you 1% of what the brain can actually do'. Why 1%? Because that is how much we know. This was his point, the brain is a super computer, a super power, something way beyond our understanding. Yes, we know a lot and we learn more and more all the time, but our brains' capabilities are far beyond what we know. Why should we look after our brain health? It really is a bit of a no brainer (pun intended). You literally cannot do anything without your brain; it impacts every single decision your body makes from voluntary to involuntary ones – it impacts your cognitive function, emotional wellbeing, physical health, longevity and productivity.

So other than the fact that we need our brain for every single thing in our body, why else is brain health important?

Maintaining brain health for longevity and performance

One of the main reasons I see people reaching out to work with me in clinic is cognitive function. Brain fog, poor memory, difficulty, concentrating and lowered mental performance are some of the biggest issues my patients face.

Patient R is a high performing executive at the top of his financial career in London. He was starting to experience increased fatigue, which he realized he had probably had for a long time, due to the stress of his job and practicalities of the travel involved. His sleep had become erratic, he was experiencing moderate brain fog some days and found he was finding it harder to be as decisive as he had been previously. He found his motivation started to wane as he found the work (which he knew he loved) harder and more stressful. He also found himself disconnecting from friends more than family. By the time he was home from work and had attended work events, his social life and connections outside of his immediate family were significantly impacted. We ran a full panel of blood work with extensive markers, a comprehensive stool test and an adrenal stress test. We found a parasite, a significant candida overgrowth, a dysbiotic bacteria, Klebsiella oxytoca and two highly elevated imbalance bacteria (chapter 11). We also found stage 3 adrenal dysfunction (chapter 9). Within the bloods which were a functional medicine blood test we could analyze systems in the body and found elevated cardiovascular risk, immune dysfunction, blood sugar dysregulation, thyroid dysfunction (sub-clinical, meaning it wasn't at disease level, but it wasn't normal either – it was below optimal function) and hormone dysfunction, with elevated oxidative stress and inflammation. In the patient's own words, he couldn't believe how much information the testing gave, over and above NHS, or even private medicals. We started working on the overgrowths in the gut while supporting the brain health and mitochondria. Within four weeks his cognitive function was vastly improved. We then turned our attention to his adrenals and sleep. By week six his fatigue, cognitive function, mental clarity and performance were all much improved. He felt like he was 'firing on all cylinders again'. We finished by working through some of things from the bloods that had become imbalanced, such as his blood sugar and cholesterol, for optimal health and longevity. What was interesting to see was that as we worked on his brain and mitochondria his thyroid function and sex hormones returned to normal – we didn't need to add extra support for these – and his inflammatory markers came back down. All of this in a relatively short space of time so that he could perform to the best of his ability and enjoy life again.

Cognitive function means involvement in memory, attention span, problem-solving abilities, language skills and decision-making. Many high performing patients with powerful and demanding jobs, at the top of their game, have told me they are struggling to make decisions like they used to. Optimizing cognitive function enhances performance, ability and your success and productivity.

As well as cognitive function, brain health impacts emotional wellbeing and mental health. The brain regulates mood, emotions and stress responses. Imbalances in brain chemistry can lead to mood disorders, such as depression and anxiety. Poor brain health can cause a person to feel less motivated at work, or disconnected from their family and friends. Promoting brain health increases emotional resilience.

Brain health is also important for physical health. Our brains communicate with every organ and system in the body via the nervous system, influencing heart rate, digestion, immune function and hormone regulation, among others.

Within the scope of optimal health for longevity, adopting healthy brain habits in midlife and beyond has proven to positively impact the risk of cognitive decline and delay the onset of age-related neurodegenerative disorders.

Neuroplasticity

One key concept that neuroscientists now understand is that your brain is changing. Early beliefs were that our brain was fixed especially in adulthood and the brain's structure and function could not be changed. Now, however, we know that the brain has a remarkable ability to change, adapt, reorganize and create new pathways. This is called neuroplasticity.

From a structural standpoint, neuroplasticity means that new synapses, the connections between neurons, can be created allowing for new neural pathways to be developed. Neuroplasticity allows for the growth of dendrites, branch like structures that receive messages from other neurons. It also allows for neurogenesis, the growth of new neurons, in some regions of the brain, especially in the hippocampus, the area involved in memory and spatial awareness.

Neuroplasticity also allows for functional changes including changes in neurotransmitter release, synaptic strength and neuronal firing. All of this increases, or decreases, neuronal activity in certain areas of the brain and impact behaviour, cognition and sensory processing.

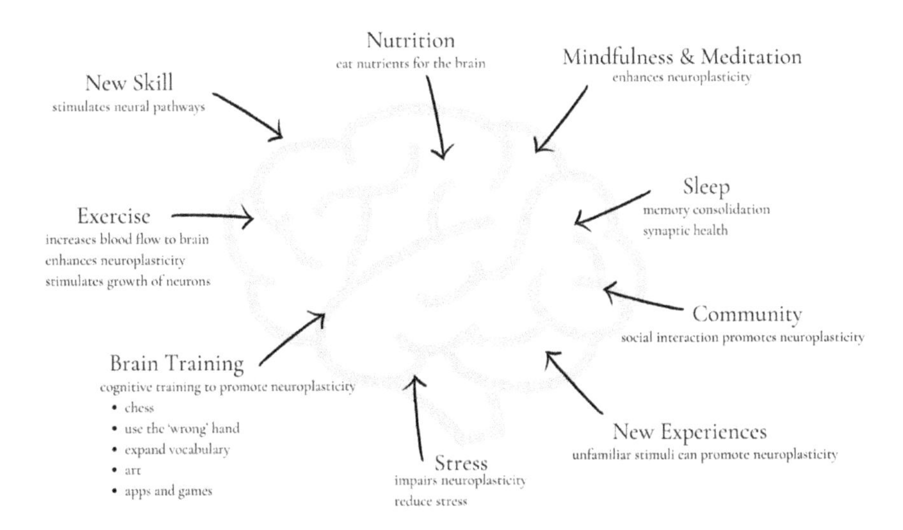

This plasticity is environment and experience dependent, meaning that our experiences, lifestyle and learning impact neuroplasticity. This means that we can influence it with the stimuli we provide it with. Think of it like this: if you went out and learnt a new musical instrument you would create new connections in the brain, in fact learning a new musical instrument creates new connections in nearly all areas of the brain. This is an example of experience influencing neuroplasticity. Our diet and lifestyle, including sleep, movement and stress, can similarly impact neuroplasticity.

All of this is positive because it means there are lots of actions we can take in the way of preventative medicine to protect our brain health.

Neuroplasticity is also key in creating adaptive responses to brain injury or damage following, for example, a stroke, traumatic brain injury or neurodegenerative disease. This means our environment and lifestyle, the choices we make, can also help us to heal.

Neurological disease

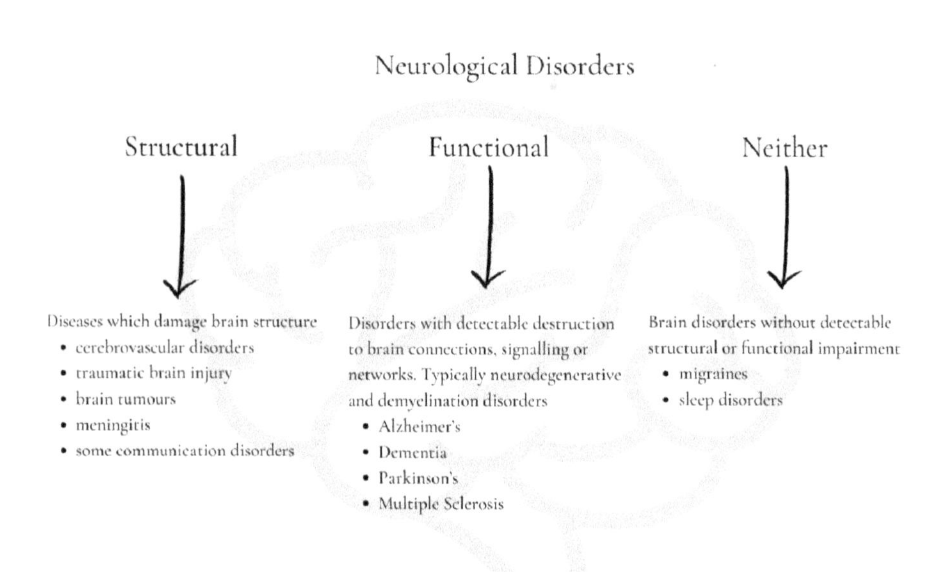

Neurological conditions, especially age-related ones, are growing as we get an ever ageing population. Lifespan over the last decades has continued to increase, but our work on healthspan has declined, our diet has become more processed, our lives and jobs have become more sedentary, our downtime has become sparse and we are up against endless low lying chronic stressors. This leads to an ever growing, ageing and infirm population. But just because our lifespan is growing doesn't mean we can't also extend our healthspan, it's just that like everything else we have to do the work to protect it.

Let's look at some of the cellular mechanisms behind neurological diseases.

Inflammation and gliosis

Inflammation in the central nervous system and gliosis are cellular responses to injury, infection or neurodegenerative processes. Inflammation activates immune cells (glial cells) in a process called gliosis, which leads to the subsequent release of pro-inflammatory molecules, cytokines and chemokines. While this process is essential for eliminating pathogens, or cellular debris, chronic or dysregulated neuroinflammation and gliosis can be a cause of neurodegeneration or neurological diseases.

Neuronal injury and cell death

A neuron will die when it reaches its limited capacity of stimuli. This cell death or damage can result in cognitive dysfunction or decline, motor decline, sensory deficits or behavioural changes.

Neurodegeneration

This is a progressive loss of neurons and/or their connections in the brain and/or spinal cord. Examples are Alzheimer's, Parkinson's and Huntingdon's. Underlying mechanisms vary but some processes are commonly present such as oxidative stress, inflammation and mitochondrial dysfunction.

Cognitive reserve vs brain reserve

First we need to understand what we mean by reserve: we are talking about the difference between the degree of damage and how that manifests clinically, in terms of symptoms or disease development. ***All of us have ageing brains.***

PEARL: From the age of 30 your brain starts shrinking – this is called brain atrophy.

But how that manifests is different in different people and this is to do with reserve.

Brain reserve and cognitive reserve can be looked at like the hardware and the software. Our brain reserve is the structure of our brain. Some people have larger brains, or more synaptic pathways, to begin with, or different volumes of grey and white matter, and therefore have a greater reserve before injury and damage will impact cognitive function – this is the hardware. When we are talking about the brain, size matters.

Cognitive reserve is the software, the brain's ability to maintain function despite age-related changes, damage or disease. It is the neuroplasticity that we can build and support. Brain-derived neurotrophic factor (BDNF) is a molecule that plays a key role in neuroplasticity, function and protection of brain health, exercise and environmental stimulation increase BDNF, so lifestyle and stimulus directly impact cognitive reserve.

Role of nutrients in brain function

PEARL: Approximately 100 billion neurons send signals via your nervous system with each neuron connecting to approximately 1000 other neurons. The adult brain has approximately 60 million neuronal connections.

Your brain is metabolically active and needs to be fed the right nutrients in order to function. There are key vitamins, minerals, amino acids and fatty acids that your brain requires as the absolute foundation in functioning at this level. In Part 3 of this book you will find all the information you need to implement a nutrition plan that supports your brain, but let's look at some of the key factors of nutrition for brain health.

Amino acids are the building blocks of proteins and they are essential for neurotransmitter synthesis, neurotransmitter receptor activation, protein synthesis, neuroprotection, cognitive function and memory.

PEARL: Your brain is 60% fat, so why would you go on a low fat diet?

It is essential that we feed the brain with the healthy sources of fats. Omega-3 fatty acids, in particular docosahexaenoic acid (DHA) and eicosapentaenoic acid (EPA), play a critical role in neuronal membrane structure and function. Adequate intake of omega-3 fatty acids is essential for maintaining neuronal integrity and supporting neuroplasticity – they have been linked to improved cognitive function, memory and mood regulation.

Antioxidants such as vitamins C and E, beta carotene and polyphenols aid in protecting the brain from oxidative stress and inflammation. Some foods such as phytonutrients have been shown to be neuroprotective, as well as supporting cognitive function.

Vitamins and minerals are essential cofactors for numerous biochemical reactions in the brain, including neurotransmitter synthesis, energy metabolism and antioxidant defence mechanisms. B vitamins, including folate, vitamin B6 and vitamin B12, are particularly important for brain health, as they play key roles in homocysteine metabolism, methylation reactions (a biochemical process to transfer atoms that supports biological functions, see page 102) and myelin (the protective covering around the nerves) synthesis. Minerals such as magnesium, zinc and iron are essential for neurotransmitter function, synaptic transmission and cognitive performance.

As important as it is to feed your brain the right nutrients, we must remove anti-nutrients such as excessive alcohol, smoking, processed foods, sugars.

Lifestyle factors and their role in brain health

You can find strategies to work on these lifestyle factors in Part 3 but let's look at the impact on brain health.

Sleep

Sleep is an essential factor in optimizing brain health as it allows the brain to consolidate memories, clear toxins and recharge for the next day. Poor sleep has been linked to cognitive impairment, worse performance, lack of mental clarity, as well as neurodegenerative disorders.

While asleep:

- New memories transfer from short- to long-term memory
- Weak or unnecessary synapses are cut
- Beneficial synaptic connections are strengthened
- The glymphatic system, a waste clearance system in the brain, becomes more active, facilitating the clearance of metabolic waste products, toxins and cellular debris accumulated during wakefulness. This is key to preventing toxicity issues. Without the glymphatic system working optimally, neurotoxins can build up.

When we are looking to optimize brain health we are often looking at improving our ability to perform better; sleep is essential for maintaining this optimal cognitive performance, including attention, concentration, problem-solving and decision-making (see case study of patient R on page 57).

Stress

Chronic stress can impair cognitive function, emotional regulation, memory formation and increase inflammation, and negatively impact brain structure and function. Chronic stress can actually lead to structural changes in the brain, particularly in regions involved in emotional regulation, memory and learning, and stress responses, such as the amygdala, hippocampus and prefrontal cortex. Prolonged exposure to stress hormones can cause dendritic atrophy (reduction in the branching of neurons), decreased neurogenesis (formation of new neurons) and alterations in synaptic plasticity (the ability of synapses to strengthen or weaken over time). When we talk about the adrenals (chapter 9) you will see how over time the brain will shut off stress hormone production in order to protect itself as cortisol can interfere with synaptic function and disrupt neural circuits involved in learning and memory processes. Additionally, chronic stress can impair prefrontal cortex function, leading to difficulties in decision-making, problem-solving and impulse control (see case study of patient R on page 57). The amygdala is particularly susceptible to stress – this is the part of the brain that deals with threat perception and emotions, including things like fear response. Chronic stress can lead to hyperactivity of the amygdala and hypertrophy (enlargement). This can lead to heightened responses and anxiety. Chronic stress also promotes neuroinflammation; the inflammatory cytokines released as part of the stress response can impair neuronal function, disrupt neurotransmitter balance and contribute to neurodegenerative processes.

Environmental toxins

Environmental toxins (chapter 18) can be detrimental to brain health and are often overlooked. Many of these environmental toxins are neurotoxic and many are things we come into contact with on a regular basis, such as pesticides on our food and the use of plastics, air pollutants and heavy metals. Exposure to neurotoxic substances can impair neurotransmitter systems, interfere with neuronal signalling and disrupt synaptic transmission, leading to cognitive deficits, behavioural changes and neurological disorders. Environmental toxins can also trigger neuroinflammation, activating microglia and the release of pro-inflammatory cytokines. They can also impair the blood-brain barrier that regulates what molecules can cross between the bloodstream and the brain. Disruption of the blood-brain barrier can allow toxins to penetrate the brain parenchyma and exert neurotoxic effects. Environmental toxins can also alter gene expression creating epigenetic changes. Exposure builds up over time as we are consistently exposed to different toxins through our lifetime.

Social connections and purpose

Studies conducted in blue zones (areas where they have found higher levels of people living to 100), such as Okinawa, Japan and Ikaria (Greece), have shown that close-knit communities and strong social networks are associated with better cognitive function, lower rates of cognitive decline and increased longevity. Social connections (chapter 16) provide emotional support, reduce stress and promote resilience, all of which are essential for maintaining brain health and cognitive vitality. Furthermore, having a sense of purpose and meaning in life is linked to improved cognitive function, better mental health and reduced risk of age-related cognitive decline and neurodegenerative diseases.

Exercise and brain function

Exercise increases blood flow and oxygenation to the brain, improving both oxygen and nutrient delivery improving cognitive performance. Exercise also stimulates the production of new neurons (neurogenesis), especially in the areas of the brain that benefit learning and memory such as the hippocampus. It promotes synaptic plasticity creating new pathways and modulates neurotransmitter balance, including serotonin, dopamine and noradrenaline (norepinephrine) which are involved in mood, stress response and cognitive function. Exercise also has anti-inflammatory and anti-oxidant capabilities, reducing oxidative stress and neuroinflammation.

The brain nutrients

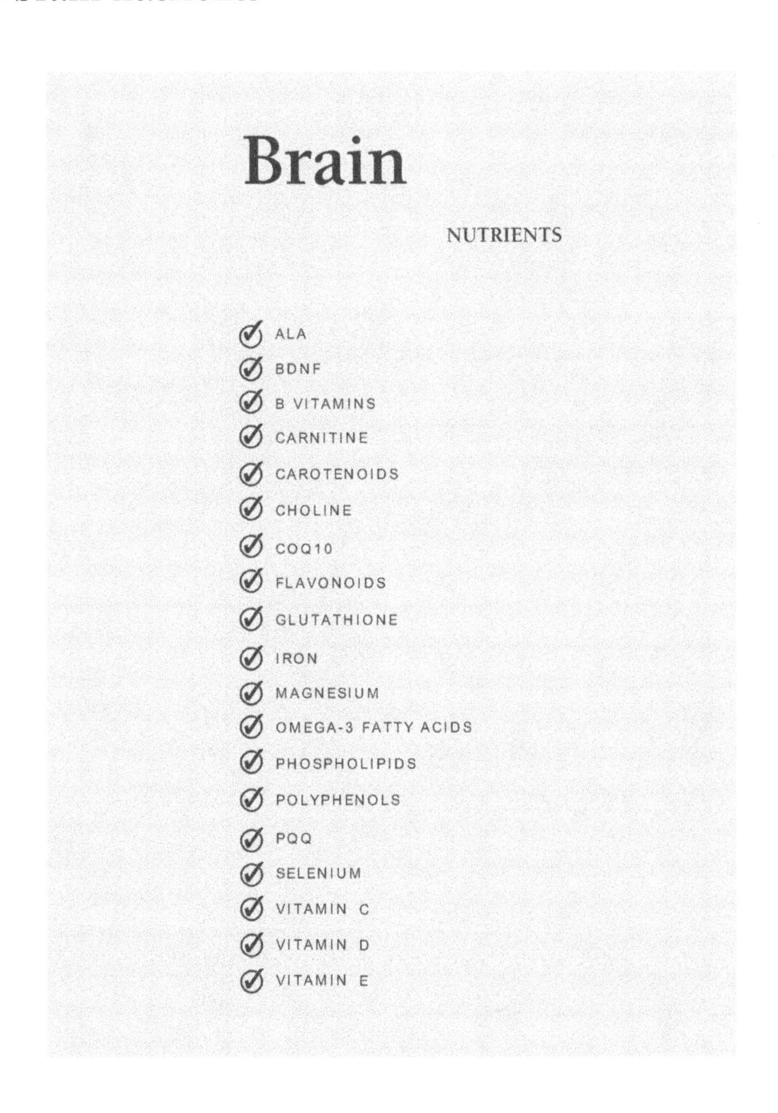

9

Adrenal health

The adrenal glands sit atop the kidneys and are pivotal in the endocrine system producing hormones that play a fundamental role in regulating various physiological functions. Cortisol, often referred to as the 'stress hormone', is released by the adrenal cortex and is integral to metabolism, blood sugar regulation and immune function. Aldosterone influences electrolyte balance and blood pressure. These hormones, collectively known as glucocorticoids and mineralocorticoids, play a vital role in the adrenals maintaining homeostasis.

The adrenals contribute significantly to the body's stress response through the hypothalamic-pituitary-adrenal (HPA) axis. This intricate system involves the hypothalamus, pituitary gland and adrenal glands, working synergistically to manage stressors. When confronted with a stressor, the HPA axis signals the release of cortisol, initiating the body's adaptive response. This response is a survival mechanism intricately linked to the autonomic nervous system, encompassing both the sympathetic and parasympathetic branches.

The autonomic nervous system

The autonomic nervous system is the part of the nervous system that controls anything that happens automatically in the body – think of all the things we don't have to consciously control such as breathing, respiration, digestion, heart rate, blood pressure. These things all go on all day, every day, without any effort from us. This is the autonomic nervous system and we have two branches of it: the sympathetic and parasympathetic branches.

Our bodies are designed to deal with stressful events, we have the 'fight or flight' response – this is the sympathetic nervous system, and for many hundreds of years it served us well to protect us when we were out hunting. The adrenal glands will release cortisol and other stress chemicals, which in turn affect the body by preparing us to either fight, or to run by increasing the output of systems in the body that would make us better at this, such as blood pressure and energy to our muscles. When the stressor is over, our levels come back down and we enter a parasympathetic state where our body focuses on rest, immune function and digestion.

However in today's society chronic low levels of stress are more common than meeting a lion while hunting for food; these low ongoing levels of stress mean that the levels don't return to normal and our sympathetic nervous system remains activated. The brain will at some point, in order to protect itself, signal for the adrenals to downregulate or switch off production of these hormones. This is because too much of these chemicals would be toxic to the brain.

What do we mean by stress?

Stress automatically leads us to think of emotional stress, but the body doesn't differentiate. Stress that impacts the adrenals could be viral load, bacteria, toxins, trauma from an injury, or surgery, emotional trauma, or day-to-day fast-paced life. It is all stress to the body and it all impacts the adrenals, and more often than not it isn't one thing, it's a buildup of different stressors, and this is where it is really important to take a full history. I run a full functional health assessment with everyone I work with, because this is how we get to the root cause.

Patient W came to see me with clear signs of HPA axis dysfunction. She was a doctor in the NHS working long hospital hours and she described herself as burnt out. She had fatigue, brain fog, difficulty concentrating, disturbed sleep and hair loss. Then she started with joint pain and muscle fatigue. We ran an adrenal stress test and found, as we expected, the flat cortisol curve, but this was just the start of her picture. From here we had to dig down and ask why this had happened to this particular person. Now there were the obvious lifestyle aspects to this case – the long hours of a doctor, lack of nutritious food while at work and nightshifts interrupting circadian rhythm – but as we picked apart her story we found she had lived in a damp building at one time, she had worked on a ward with contact with infectious diseases, she had had Epstein Barr Virus (EBV) when she was younger and she had had a severe gastric upset while travelling abroad in her 20s. So while we were looking at burnout, asking why and getting to the root cause led us to run mycotoxin testing, viral tests and a comprehensive stool test. We found reactivated EBV, HHV-6, parasitic infection and two mould toxicities. All of these act as stress on the body. As we worked on each of these one by one, the adrenals started to balance, symptoms abated and the body found homeostasis again. This is where functional medicine comes into its own; we can't just look at one part of the body, it is all interconnected.

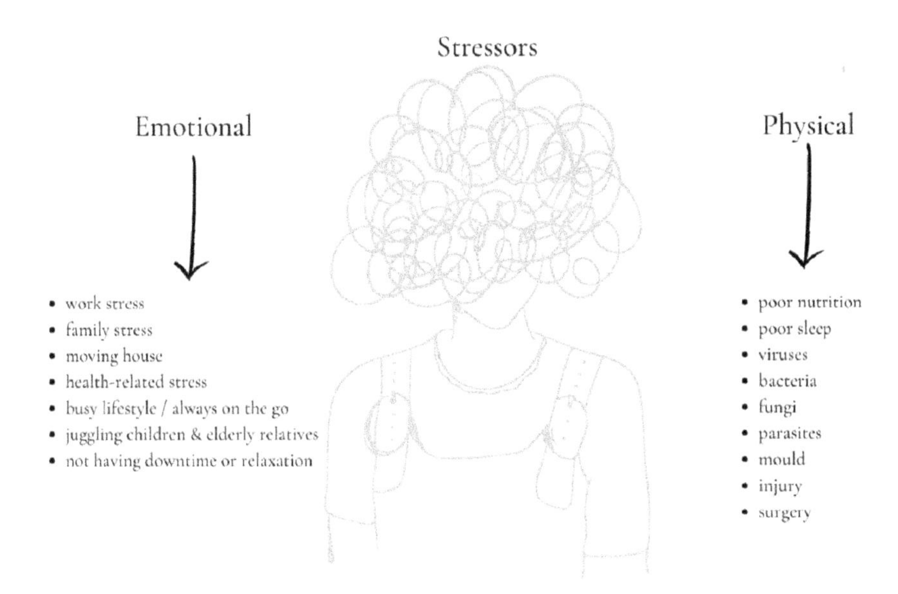

Why it's important to understand the adrenals

Adrenal health is a core pillar of The MitoImmune Method for good reason – the adrenals exert a profound influence on various body systems, highlighting the interconnectedness of the pillars and the importance of taking a whole body approach. The adrenals are usually most talked about when we discuss energy, however they secrete many hormones involved in processes such as energy production, fluid and electrolyte balance, fat storage, sex hormone production, blood sugar, blood pressure and more.

Adrenal–mitochondria connection

Cortisol released by the adrenals can impact mitochondrial function and therefore our energy. Initially acute cortisol can actually enhance mitochondrial action as a response to our fight or flight response (more on this on page 67) to an immediate stressor, however chronically elevated levels of stress can disrupt mitochondrial function. Elevated cortisol can also promote oxidative stress impairing mitochondrial function. The adrenals impact metabolism and energy utilization contributing to energy balance, while mitochondria maintain energy homeostasis at a cellular level.

Adrenal and mitochondrial dysfunction has been linked to chronic fatigue and other autoimmune conditions.

Adrenal–immune connection

The stress hormone cortisol, produced by the adrenals, plays a pivotal role in modulating the immune response. While acute stress can enhance immune function temporarily, chronic stress and sustained high cortisol levels may lead to immune suppression, making the body more susceptible to infections and development of inflammatory, autoimmune and chronic health conditions.

Adrenal–gut connection

Elevated cortisol from chronic stress can lead to changes in gut motility, blood flow and even the balance of our gut microbiome (chapter 11). Through the gut–brain axis, which is bidirectional, the nervous system, hormone balance and immune system can all be impacted. Adrenal hormones can also impact secretion of stomach acid and the contraction of smooth muscle in the gut impacting our digestion and leading to symptoms such as indigestion, bloating and bowel disruption. Prolonged elevated cortisol levels can also contribute to the loosening of tight junctions in the intestinal wall. With the changes to gut motility and intestinal permeability, the gut microbiota can alter as the environment within the gut changes. Where the permeability of the intestinal lining is compromised, inflammation can occur and this is a known trigger for autoimmune diseases.

Adrenal–brain connection – the HPA axis

We've looked at the impact of stress on the brain in chapter 8. Here I want to look further at the impact of neurotransmitters.

Cortisol and neurotransmitters

- **GABA (Gamma-Aminobutyric Acid):** Cortisol can influence the balance of neurotransmitters in the brain, and it has inhibitory effects on GABA, an important neurotransmitter with calming and anti-anxiety properties.
- **Serotonin:** Cortisol can impact serotonin levels, a neurotransmitter associated with mood regulation, potentially contributing to mood disorders such as depression.

- **Glutamate:** Cortisol can modulate the activity of glutamate, an excitatory neurotransmitter involved in learning and memory, leading to conditions like anxiety and cognitive dysfunction.

Adrenaline (Epinephrine) and neurotransmitters

- **Adrenaline (Epinephrine) and Noradrenaline (Norepinephrine):** Both adrenaline and noradrenaline are released from the adrenal medulla in response to stress. These neurotransmitters enhance alertness, increase heart rate and prepare the body for a 'fight or flight' response.
- **Dopamine:** The release of adrenaline can influence the release of dopamine, another neurotransmitter associated with motivation and pleasure.

Adrenal–hormone connection

The adrenal glands play a significant role in influencing sex hormones, although they are not the primary glands responsible for sex hormone production. The interplay between the adrenal glands and sex hormones occurs through the production of certain precursor hormones that can be converted into sex hormones. Dehydroepiandrosterone (DHEA), for example, is a precursor to both male and female sex hormones, androgens and oestrogens; it is converted into hormones such as testosterone and oestrogen in various tissues in the body.

The 'cortisol steal' occurs when cortisol production, due to stressors on the body, is prioritized over other essential hormones, potentially disrupting hormonal balance. This happens because the production of all of our hormones is essentially the same – it requires a form of cholesterol called low-density lipoprotein (LDL) cholesterol. As we only have a certain amount of it available we can only make a certain amount of hormones. So as we produce more and more cortisol we run the risk of producing less of our other hormones such as oestrogen, testosterone or progesterone, so our adrenals are heavily interconnected with our hormone health. We can therefore see the development of other symptoms not necessarily associated with HPA axis dysfunction such as low libido and sex drive due to low testosterone, impaired muscle strength, further drops in energy and increased risk of chronic illness and autoimmune disease. Additionally the adrenal glands play a role in weight loss by regulating metabolism, energy levels and appetite. Stress can cause the body to produce cortisol, which can increase appetite and cause the body to hold onto fat stores.

Adrenal–thyroid connection

The adrenals and the thyroid produce hormones as part of the endocrine system that impacts metabolism and energy production. Cortisol from the adrenals influences metabolism, and thyroxine (T4) and Triiodothyronine (T3) regulate the body's metabolic rate. The adrenals' stress response can impact thyroid function and suppress conversion of T4 to T3 which impacts metabolic rate. The 'cortisol steal' can also lead to imbalanced thyroid hormones. The adrenals and thyroid can impact energy balance leading to fatigue, weight fluctuations and energy levels.

Symptoms of adrenal dysfunction

Dysregulation of the adrenals carries profound implications for health, manifesting in symptoms.

Symptoms of adrenal dysfunction

- Fatigue
- Brain fog
- Tired but wired (a second wind in the evening)
- Insomnia
- Headaches
- Palpitations
- Dizzy spells (especially when standing)
- Poor circulation
- Decreased tolerance to cold/always feeling cold
- Joint aches and pains
- Muscle weakness
- Low immunity
- Craving salt or possibly sugar
- Gut distress (stomach aches/constipation/diarrhoea)
- Loss of interest or motivation
- Low mood
- Withdrawing from work, family and/or friends
- Anxiety
- Frustration

PEARL: People struggling with chronic daily stress and burnout cost $1 trillion of lost productivity globally every year.

Does adrenal fatigue exist?

You may have heard doctors say adrenal fatigue doesn't exist, and here's the thing, they are right. Why? Because it's a fashionable terminology that has been used on social media and in the public domain, but it's actually incorrect, and this is where the confusion has occurred.

You will still see practitioners say adrenal fatigue – I do sometimes. Why? Because the public, you, understand what I mean by it. If I talk about HPA axis dysfunction, many people don't know what I mean.

So why is it wrong? Well, the adrenals don't actually get fatigued. What actually happens is that the communication pathway (hypothalamus–pituitary–adrenals) suffers dysfunction. So it is actually HPA axis dysfunction – this is also important as it acknowledges the involvement of the brain.

The cortisol curve

Cortisol and melatonin are hormones that help control your circadian rhythm – this is the wake–sleep cycle of your body. They should act inversely to each other. Cortisol is high in the morning to help us get up and moving, while melatonin is low. Melatonin is high in the evening to promote sleep, while cortisol is low.

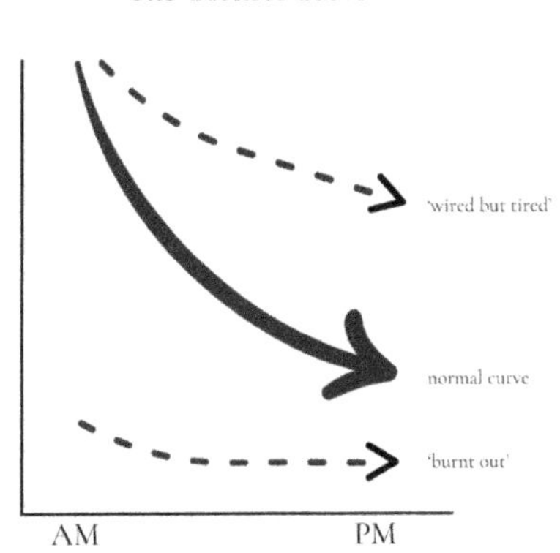

The cortisol curve is a reflection of these cortisol levels throughout the day, peaking early morning and gradually decreasing over the course of the day, before rising again in the early hours, peaking on waking. Chronic stress can alter this curve, resulting in an elevated or flattened pattern, depending on the stage of dysfunction.

Chronic stress disrupts this delicate balance, leading to dysregulation of the adrenals. Prolonged stress initially triggers a sustained elevation of cortisol levels, impacting the adrenal glands and potentially leading to adrenal dysfunction. As your body gets used to elevated levels of cortisol, more is required in order to have an influence so levels keep rising. However, too much cortisol can be toxic to the brain and so eventually the brain will intercept and shut off production of the stress hormones to protect you. Your brain will always keep you alive by protecting itself and deciding which functions you need to focus on and protect at any given time. When this happens the cortisol levels drop and you crash.

The stages of HPA axis dysfunction

There are different stages of adrenal dysfunction and knowing where a person is at can really help us to optimize their health. What we are really asking is how much resiliency does a person have left? ***It's a bit like having some money in a savings account for a rainy day. How much of that does a person have? This is your resiliency.***

Stage 1 – the natural stress response

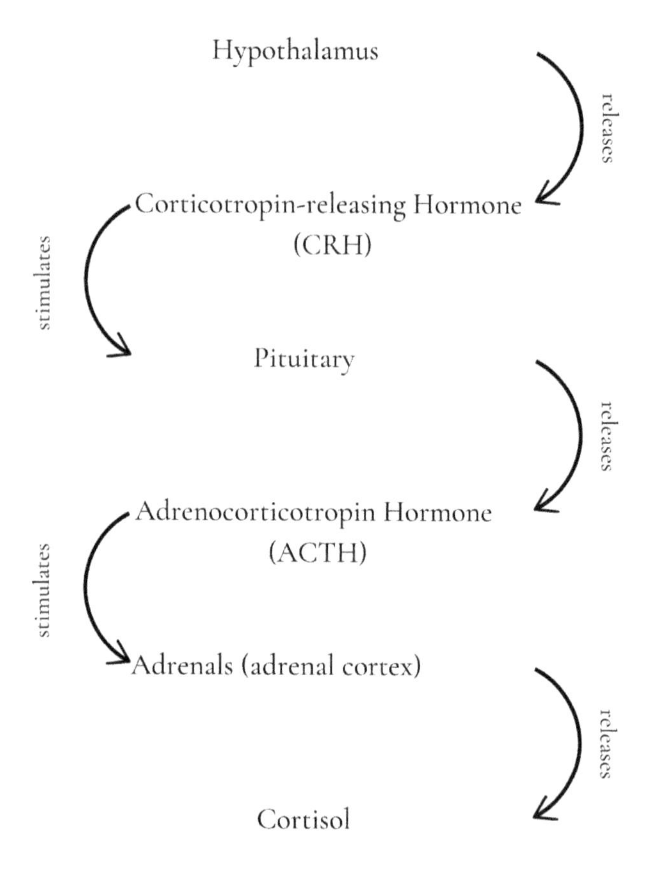

When needed you are in your fight or flight response – it is your immediate reaction to a stressor. Your cortisol is high and your DHEA is either normal or high. Your hypothalamus is activated to upregulate things like heart rate, respiration and temperature, and downregualte things such as digestion, libido and hunger. This is preparing you for danger. A cascade of events is activated and importantly brought back to balance quickly as soon as the threat is over.

This is a life-saving system, and in danger you want this to happen, so a short-term alarm phase isn't a concern; it is when it becomes chronic that we have to worry.

Stage 2 – the wired then tired phase

If chronic stress continues, DHEA starts to drop and cortisol will still be elevated. Your resiliency is starting to be impacted. And sometimes at this stage people actually feel quite good – this is the wired phase; you may be able to work late and feel super focused, but people don't always realize they are being negatively impacted.

Stage 3 – crashed

However, as you push through and push your body, your cortisol levels start to drop. This is where your cortisol levels have been so high that it was becoming toxic to your brain and the brain has protected itself by stopping cortisol production.

In summary

So you go from feeling really good with lots of adrenaline and cortisol and getting loads done and being able to work really long hours, in fact probably feeling a little bit like a super hero, to feeling wired but tired, to feeling completely exhausted and burnt out.

The adrenal nutrients

Adrenal

NUTRIENTS

- ADAPTOGENS
- ALA
- B VITAMINS
- MAGNESIUM
- OMEGA-3 FATTY ACIDS
- VITAMIN C
- VITAMIN E
- ZINC

10

Thyroid health

What is the thyroid?

The thyroid is a butterfly shaped gland that is an endocrine organ, meaning it produces hormones, and it sits at the lower front part of your neck, with the butterfly wings (the lobes) sitting on either side of the trachea connected by a band of tissue known as the isthmus. It plays a crucial role in regulating numerous bodily functions through the hormones it produces.

The thyroid is an essential part of our endocrine system. It is involved in many physiological functions, but primarily it produces and secretes hormones that influence energy balance, metabolism and cellular activity.

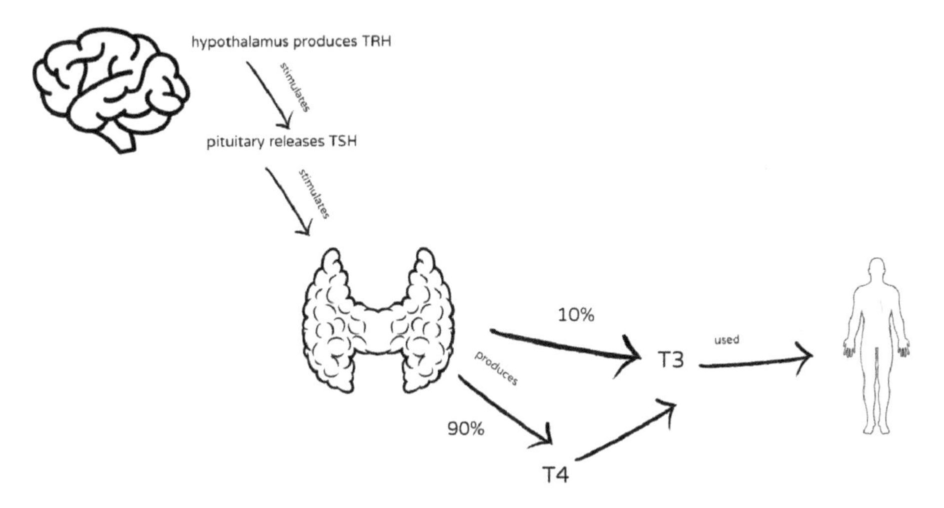

The two main hormones produced by the thyroid are thyroxine (T4) and triiodothyronine (T3). This occurs via the hypothalamus–pituitary–thyroid (HPT) axis. The hypothalamus produces thyrotropin releasing hormone (TRH), which signals for the pituitary to release thyroid-stimulating hormone (TSH). Many people think TSH is a thyroid hormone as it is the most common marker you will see measured when you have thyroid blood tests done, but it is actually coming from the brain, the pituitary. TSH then stimulates the thyroid to produce the thyroid hormones T4 and T3 – 90% of the hormones the thyroid produces are in the form T4 and only 10% are T3, however T4 is the inactive form and T3 the active form. So your T4 must be converted to T3 for it to be used by the body.

There is a very narrow window for fluctuation in the thyroid hormones and this process allows for homeostasis to be upheld.

Why is it important to balance your thyroid for optimal health?

The thyroid influences the function of virtually every cell in the body, therefore thyroid balance is essential to cellular health.

The thyroid hormones' involvement in metabolism includes regulating energy production and consumption. The thyroid hormones facilitate the conversion of nutrients from food into energy (chapter 6 and more below). Adequate thyroid hormone levels are also essential for optimal cognitive function, memory, concentration and mood regulation. Sub-optimal thyroid function cannot only lead to the development of chronic and autoimmune thyroid conditions (more on this on page 81), but thyroid dysfunction can impair the body's ability to mount an effective immune response.

The thyroid connections

Thyroid and energy

As we saw in chapter 6, the thyroid is heavily involved with the mitochondria in energy production at the cellular level and mitochondrial biogenesis (that process of increasing numbers of mitochondria). Imbalances in thyroid function can lead to fatigue and decreased energy levels, affecting overall vitality and quality of life. We also saw that thyroid hormones influence mitochondrial ROS production (page 43) by modulating activity and anti-oxidant defence mechanisms. Imbalances in thyroid function can lead to either excessive ROS production or impaired anti-oxidant capacity, contributing to oxidative stress and mitochondrial dysfunction. Finally we saw that thyroid hormones regulate the mtDNA. T3 directly binds to mitochondrial receptors and stimulates mtDNA transcription, leading to increased expression of mitochondrial genes involved in energy metabolism.

Thyroid and immune function

Thyroid conditions commonly involve the immune system, and the chronic inflammation associated with autoimmune thyroid conditions can further dysregulate immune function – we look at thyroid conditions on page 81. Thyroid hormones also play a modulatory role in immune function by influencing the activity of immune cells and cytokine production (chapter 7). Thyroid dysfunction can impact the balance of pro-inflammatory and anti-inflammatory cytokines, leading to immune dysfunction, and also impacts your body's ability to mount an effective immune response, making you more susceptible to infections. Disruption to the antioxidants in your body can lead to increased inflammation and immune dysfunction.

Thyroid dysfunction has also been linked to thymic atrophy – the thymus being a gland that makes T-lymphocytes (a type of white blood cell). T-lymphocytes, you may recall from chapter 7, are essential to your adaptive immunity and therefore your immune memory.

The thyroid and the brain

We have already discussed the HPT axis and the involvement of the brain in signalling thyroid hormone production, but the thyroid is also interconnected with the brain in other ways too. Thyroid hormones play a crucial role in brain development and function, particularly during foetal development and early childhood. They are also essential for maintaining cognitive function throughout life. Adequate thyroid hormone levels are necessary for optimal neurotransmitter synthesis, synaptic transmission and neuronal plasticity in various brain regions involved in learning and memory. The thyroid–brain connection also extends to mood regulation, with thyroid hormones influencing neurotransmitters such as serotonin, dopamine and noradrenaline (norepinephrine), which play key roles in mood regulation.

We also commonly see symptoms such as poor balance in those with Hashimoto's (more on this on page 81), this is due to cerebellar dysfunction that can occur, resulting in symptoms such as ataxia, tremors and impaired coordination. The cerebellum, a brain region involved in motor control and coordination, contains thyroid hormone receptors and is sensitive to thyroid hormone levels.

Thyroid and the adrenals

As we saw in chapter 9, the adrenals and the thyroid are part of the same endocrine system. Thyroid hormones influence the adrenals and this can impact the production of our stress hormones. Conversely, chronic stress impacts adrenal function which can affect thyroid function.

Thyroid and the gut

The thyroid hormones have a direct effect on the motility of your gut, the contraction and relaxation of smooth muscle along your digestive tract, which ensures food moves through your system as it should. This is why in hypothyroidism we often see symptoms like constipation, bloating and stomach cramps. Conversely, we can see diarrhoea in hyperthyroid patients.

The thyroid hormones also help regulate the secretion of digestive enzymes which helps break down and absorb the food we eat, impacts our gut microbiome and helps maintain the integrity of the tight junctions in the intestinal wall, keeping our gut barrier healthy and protecting us from intestinal permeability (chapter 11). Due to the thyroid's involvement in neurotransmitters it is now also thought that the thyroid may impact the gut–brain axis.

Thyroid and other hormones

Thyroid imbalances can disrupt menstrual cycles, impair fertility and increase the risk of complications during pregnancy, such as miscarriage, preterm birth and developmental abnormalities in the foetus. Thyroid hormones also influence bone metabolism by regulating the activity of osteoblasts and osteoclasts, the cells responsible for bone formation and resorption, respectively.

The thyroid gland is involved in modulating our hunger hormones which regulate our appetite and signal us when we are full. These hormones, leptin and ghrelin, can be altered by thyroid hormone levels.

Thyroid and its role in cardiometabolic function

A properly functioning thyroid gland ensures that metabolic processes operate efficiently, helping to maintain a healthy weight and prevent metabolic disorders such as obesity, blood sugar imbalance, insulin resistance and metabolic syndrome. Thyroid hormones help regulate body temperature by influencing metabolic rate and heat production. They also have significant effects on the cardiovascular system, including heart rate and contractility. Imbalances in thyroid function can lead to changes in heart rhythm, increased risk of hypertension and alterations in lipid metabolism, all of which can contribute to cardiovascular disease if left untreated.

Toxins and the thyroid

The thyroid is highly susceptible to environmental toxins and chemicals that disrupt the thyroid. Many different toxins have been found to impact the thyroid.

You will find more on toxins in chapters 18 and 19, including how to reduce your toxic exposure and brands of supplements I love to help you do this.

Thyroid conditions

Thyroid conditions are one of the fastest growing autoimmune conditions.

PEARL: One in eight women will develop a thyroid imbalance during her lifetime.

Hypothyroidism occurs when the thyroid gland produces insufficient amounts of thyroid hormones; on the other hand, hyperthyroidism results from an overproduction of thyroid hormones. These thyroid conditions can be autoimmune in nature; in fact in both cases it is more likely that you have the autoimmune condition than simply hypo- or hyperthyroidism, with Hashimoto's being autoimmune hypothyroiditis and Grave's being autoimmune hyperthyroiditis.

PEARL: Approximately 90% of hypothyroidism cases are Hashimoto's – an autoimmune disorder.

Symptoms

Hashimoto's (Hypothyroid)	**Grave's (Hyperthyroid)**
Fatigue	Fatigue
Weight gain or difficulty losing weight	Weight loss
Cold intolerance	Heat intolerance
Brain fog	Sweating
Problems with memory or concentration	Blurred vision
Hair loss, or dry hair and skin	Pain in eyes
Balance issues	Bulging eyes
Constipation	Diarrhoea
Muscle weakness, aches or pains	Muscle weakness or Tremors
Joint pain	Redness or thickening of the skin
Depression	Anxiety
Menstrual irregularities	Insomnia
Bradycardia	Tachycardia/Palpitations
Goiter	Goiter

It's important to remember that hormones fluctuate and therefore you can have flares of the alternate thyroid condition, for example if you have hypothyroidism you may experience hyper flares and therefore see the symptoms of hyperthyroidism at those times. You may also not get all of these symptoms, you may only have a few.

Thyroxine is commonly given as a medication to treat thyroid conditions and this can be an important first step to balancing TSH. However there are three different areas in the pathways where thyroid function can be impaired: the TSH (balanced with medication), the conversion of T4 to T3 and the uptake of T3 into your cells (see the image on page 77). On top of this, if you are only taking medication for your autoimmune thyroid condition you are ignoring a whole important part of the picture – as well as working on thyroid function you need to work on balancing the autoimmunity and immune function (chapter 7).

Further considerations with autoimmune thyroid conditions

If you have autoimmune thyroiditis, such as Hashimoto's, there are some further considerations you may want to consider. You can make these adaptations to The MitoImmune Plan in Part 3.

Gluten

As we've already seen in chapter 7, with any autoimmunity it is common that intestinal permeability is a root cause. With autoimmunity such as Hashimoto's or Grave's, the last thing we want is a constantly activated immune system (read more in chapter 11). In thyroid patients, a gluten-free diet has been shown in the studies to reduce TPOAb and TGAb levels (the thyroid antibodies that elevate, indicating Hashimoto's), increase vitamin D levels, improve inflammation levels, lead to less immune reactivity and allow a decrease in thyroid medication. Also, the vast majority of Hashimoto's patients have been found to have at least one of the HLA-DQ genes – these are the genes that cause Coeliac disease, and having one means the person will at least be gluten sensitive and at worst Coeliac. If Coeliac, there will be a huge improvement in symptoms, but even if just sensitive to gluten, inflammation in the body will decrease which is key.

Dairy

Molecular mimicry is a phenomenon in which the immune system, triggered by exposure to certain foreign substances, may mistakenly target similar-looking structures on the body's own cells. This can lead to autoimmune reactions, where the immune system attacks its own tissues. While molecular mimicry is a complex and multifaceted concept, some studies and hypotheses have explored the potential role of dairy proteins in triggering autoimmune responses. Casein is a protein found in milk, and molecular mimicry may occur between casein and certain tissues in the body. A study in *Autoimmunity Reviews* showed molecular mimicry between a sequence of the beta-casein protein and a component of the thyroid gland. Therefore reducing dairy can help to prevent this molecular mimicry mechanism of autoimmunity. Where possible swap dairy products for non-dairy alternatives, such as coconut or almond milk and yoghurt.

Iodine and Hashimoto's

Iodine – there is a mass of research against using iodine in Hashimoto's patients, however some practitioners still promote its use. This is extremely concerning. Hashimoto's is not caused by an iodine deficiency; instead your follicular cells are being destroyed by autoimmune processes. Iodine actually flares up this process in Hashimoto's patients. A low iodine diet is recommended and where antibodies remain stubbornly high an iodine restriction diet can be beneficial. Some people initially feel better taking iodine, or eating iodine, which is why some practitioners will still recommend it, but what is actually happening is that as you have the iodine, initially more thyroid hormones are released and in circulation (which can show some signs of improvement), but then over time it causes more follicular cell destruction and you feel worse, with more damage that is permanent. If you feel you are trying everything but nothing is improving, a restricted iodine diet can be tried. I have often had cases whereby I remove a Hashimoto's patient from a multi-supplement which contained iodine to find they improve.

Goitrogens

Goitrogens are linked to the thyroid and to iodine. They are defined as substances that disrupt the production of thyroid hormones by interfering with the uptake of iodine in the thyroid gland. What happens then is the pituitary in the brain releases TSH and promotes the growth of thyroid tissue, which can lead to a goiter. Now what is important to note is that there are food goitrogens and chemical goitrogens, and the two need to be considered as entirely separate things. So let's take food goitrogens first. Again this is an area where you will find practitioners still recommending that you avoid food goitrogens – what has happened here is that some studies were done in a test tube and these practitioners are using the results of those. But the latest research has been done in humans and this shows that the same problems of inhibiting iodine uptake into cells doesn't occur in the human body, and actually there are huge benefits to thyroid patients from the nutrients in goitrogens. Goitrogenic foods are things like cruciferous vegetables, soy, cassava, lima beans, lentils, sauerkraut, sweet potato, cabbage and kale. One study actually tested a broccoli drink (broccoli is a goitrogen) and found no impact in the human body to iodine uptake, but huge benefits to inflammation in the body. Food goitrogens actually increase glutathione levels. Elevating glutathione is important in thyroid patients to protect the thyroid from damage (remember it's that master antioxidant as we saw on pages 39 and 145), and also in patients with Hashimoto's. So you want to consume goitrogens as foods. Chemical goitrogens ought to be avoided. (Read more about these in chapters 17 and 18.)

Lectins

Lectins are found in nuts, seeds and grains. Lectins can cause cross reactivity in people, especially those with autoimmunity. One of the things lectins can cross react with is TPOAbs. The antibodies made against lectins can bind to TPO in the thyroid gland. Now one component of gluten is wheat germ agglutinin which is the lectin part of these foods. This is why some people will say I got tested for gluten but I was ok – they may not have tested for wheat germ agglutinin and this is why we use autoantibody testing against food at Goode Health instead of your standard food allergy testing, which is extremely limited in how much you can use the results. We can also test you against lectin antibodies. We do this as it is possible you don't appear to react to gluten, or even show up a reaction on a test, but lectins could be pushing up your antibodies and autoimmune reactivity.

We want to test and find out for each individual what they should and shouldn't be eating, as we don't want to remove any more foods from the diet than necessary, because this leads nutrient deficiencies and dysbiosis (chapter 11) which is also a trigger for autoimmunity.

Patient Y, who was diagnosed with Hashimoto's, came to me having tried every diet imaginable: paleo, autoimmune paleo, keto, elimination, carnivore, you name it she had tried it. Initially she felt better as she eliminated foods from her diet, then, in her words, she said 'everything stopped working'. She had plateaued with her health and then she actually started to decline. This is something I see a lot in people – the utter frustration at doing everything they read about and not seeing improvements. I knew looking at her food journal that I was going to be reintroducing foods into her diet; we also did a comprehensive stool test and some antibody food testing (chapter 19). We found dysbiosis in the gut and bacterial overgrowths, nothing pathogenic, but not enough of the good stuff. I told her we were going to start adding foods in, which she was quite anxious about as she was expecting to see further decline in her health, because originally cutting out foods had helped. We did the reintroduction in a very structured way that allowed us to monitor her reactions (chapter 11). We started with foods I suspected would be best tolerated while we waited on her autoantibody food testing. Very quickly we saw improvements in her health – she had more energy, her brain fog lifted, her memory improved. When we got the autoantibody testing back we did find that gluten and wheat germ agglutinin were indicated. We left gluten and lectins out of the diet but reintroduced everything else, via the appropriate method to double check everything, and her health significantly improved. Now that she was feeling much improved we could start digging for root causes to further benefit her health.

Nightshades and sodium (salt)

Nightshades and sodium are also important for thyroid health, especially autoimmune thyroid conditions – we discussed these in chapter 7 if you want to read more.

Thyroid testing

You can learn more about thyroid testing in the functional testing chapter (chapter 19), but what is important to understand here is why functional testing of the thyroid becomes important. Within conventional medicine, commonly only TSH is checked and monitored; you may be able to get T4 or T3, but many people that I see can't. The problem with just checking TSH is that it gives a very small part of the picture. We also want to see how you are converting from T4 to T3. We can also look at other markers outside of thyroid function that are complimentary.

Many people I see also don't know whether they have hypothyroidism or Hashimoto's, or hyperthyroidism or Grave's; sometimes because they have never had their antibodies tested via conventional medicine and sometimes they have had it done once, but haven't been told which they have. It's really important to know if you also need to work on autoimmunity. If we think back to the immune health chapter (chapter 7), those antibodies show up a long time before thyroid function will be impaired, so they are an important part of the preventative medicine picture.

Really important to note is you can't compare your antibody levels to someone else's. Often I find people will say something like, 'my TPO antibodies are 900 and my friend's are only 90'. We have to look at the bigger picture. The antibodies are what tag the tissue but they aren't what is causing the damage; how your immune system is functioning is important here. If you had TPO antibodies at 900 but your immune system is balanced and not coming out to mop up that tissue, you will have little damage to the thyroid gland; if you had TPO antibodies of 90 but your immune system is in total overdrive and all of that tagged tissue is going to get destroyed, you are worse off at 90 than you are with the 900. Now this is a simplification but you get the idea. Don't compare to other people, compare to your previous and future results to look at progression, remission, relapses and flares.

In the online book resources section on the website you will also find a free guide to help you understand, track and manage your thyroid test results. You can also order a full thyroid panel, with antibodies, with a home test if you are struggling to get a full thyroid panel done.

The thyroid nutrients

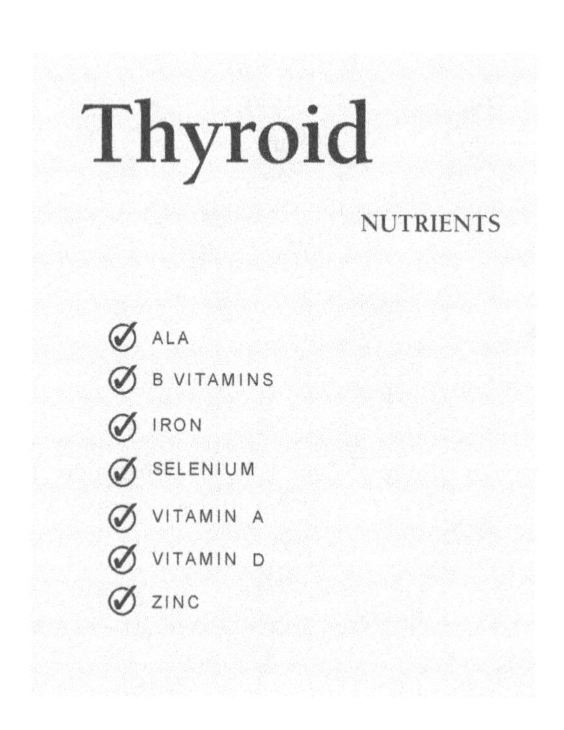

11

Gut health

Your gut microbiome

The gut microbiome is an ecosystem of trillions of microorganisms, including bacteria, viruses, fungi, parasites and other microbes. They all reside in the digestive tract and this intricate ecosystem plays a vital role in overall health and wellbeing. Imagine the rainforest with all the species that live in it, in balance, in harmony creating an ecosystem that allows for all species to succeed. This is your gut microbiome, or at least this is what a healthy microbiome should be.

You may also hear it referred to as the microbiota and sometimes people use these terms interchangeably. The microbiota refers to the actual organisms themselves, while the microbiome encompasses the microorganisms plus the whole of their genetic material. There are trillions of bacteria alone living in our mouths, on our skin and in our gut.

So just how important is the gut microbiome? Let's look at it this way...

PEARL: The human genome carries around 22,000 genes, while researchers estimate that the microbiome in a human carries 8 million unique genes. In other words, you have around 360 times more bacterial genes in your body than you do human genes.

We really are more bacteria than human. Now do you think it's important to look after your gut microbiome?

Your microbiome is dynamic, it changes over time, from developing in early childhood to fluctuating with environmental impacts as we grow. Achieving a balanced and diverse microbiome is key to optimal health. Environmental factors – nutrition, lifestyle, toxins and infections – have the biggest impact.

PEARL: Did you know your gut microbiome is unique to you just like a fingerprint?

This is where personalized medicine (chapter 4) becomes key to success. We can now assess your microbiome, gut health, immune function and more with a comprehensive stool test (chapter 19) which I run with almost everyone I work with.

Why is gut health important?

We want as much diversity in the gut as possible, with as little overgrowths as possible. Your microbiome has a profound impact on digestion, nutrition absorption, metabolism, energy, hormones, the thyroid and even plays a crucial role in regulating your immune system.

The microbes in the gut help to break down complex carbohydrates, fibres and other compounds that our body alone cannot digest. This produces short chain fatty acids and other by-products which contribute to overall gut health and influence metabolic processes. Certain bacteria found in the microbiome play a role in nutrient absorption; if the microbiome is imbalanced it can lead to malabsorption of some nutrients and deficiencies, which in turn can impact various functions of the body.

Gut-immune axis

PEARL: Over 70% of your immune system resides in the gut

The microbiome influences our immune system development and function, with a healthy microbiome supporting a well regulated immune response, while an imbalanced microbiome is associated with immune disorders and chronic inflammatory conditions. Often referred to as the gut-immune axis, there is a dynamic relationship at play between the gut and our immune system that has a pivotal role in maintaining overall health. At the forefront of the gut's immune defence lies secretory immunoglobulin A (sIgA), that serves as the body's first line of defence against invading pathogens. Produced by plasma cells in the mucosal lining of the gut, sIgA plays a crucial role in preventing the attachment of harmful microorganisms to the intestinal lining. This immunoglobulin acts as a shield, neutralizing and preventing the entry of pathogens into the bloodstream.

The microbiome plays a crucial role in training the immune system during its development. Our immune cells gain their education by being exposed to various microbes – through this exposure they learn to distinguish between harmless substances and potential threats. They can determine the difference between self- and non-self-tissue. This immunological education is vital for the establishment of immune tolerance and the prevention of unnecessary inflammatory responses.

Remember loss of immune tolerance is essentially what causes autoimmune diseases to develop.

Anthony Haynes, a renowned clinical functional medicine practitioner, told me the following when he came on my podcast.[2]

PEARL: More immune decisions are made by your gut in one day than in your whole body in your whole life.

Microbes in the gut ferment dietary fibres to produce short-chain fatty acids (SCFAs), such as butyrate, acetate and propionate. SCFAs play a significant role in modulating immune responses, promoting anti-inflammatory signals and contributing to the maintenance of a balanced immune system. Outside of the microbiome our stomach acid in the gut, which is impacted by the health of our gut, acts as a formidable barrier, killing harmful bacteria and parasites before they can reach the intestines.

Gut–mitochondria connection

In its simplest form the gut is linked to the mitochondria via nutrient absorption. Your gut is responsible for absorbing nutrients from the food that you eat – these nutrients are used by the mitochondria to produce energy (chapter 6). A healthy gut will better absorb nutrients while a gut with an imbalance or dysfunction can contribute to mitochondrial dysfunction. The gut microbiome also influences mitochondrial biogenesis, the process by which new mitochondria are generated within cells.

SCFAs, especially butyrate, have been shown to enhance mitochondrial function. Studies have shown that butyrate positively impacts neuroinflammation and oxidative stress in the brain positively impacting our mitochondria. It has also been found that butyrate induced mitochondrial function can positively impact the health of our intestine wall, impacting gastrointestinal conditions and diseases.

Gut-brain axis

The microbiome in your gut communicates bidirectionally with your brain and nervous system. The enteric nervous system (ENS) is part of your autonomic nervous system and is often referred to as the 'second brain'. It is a complex network of over 100 million neurons that regulate the functioning of the gastrointestinal (GI) tract, containing a vast array of neuronal cell types, including sensory neurons, interneurons and motor neurons.

---

[2] www.nicolegoodehealth.com/the-goode-health-podcast/the-hidden-role-of-infectious-agents-in-autoimmune-diseases-with-antony-haynes

What is so unique about this is that the ENS operates independently of the central nervous system (CNS), but maintains communication with it through various neural connections. The GI tract is the only part of our body that has evolved to have its own nervous system that can act independently to the brain in this way. Instead the brain signals the ENS and the ENS communicates back to the brain, they interact to regulate functions such as motility, secretion, absorption and local blood flow, allowing for efficient digestion and nutrient absorption.

Your vagus nerve is the nerve that runs from your brainstem to your gut, with the messaging signals going both ways. This can impact mood, behaviour and cognitive function and performance. This connection is so strong that there are people who have coeliac disease – a gut autoimmune disorder – who experience no digestive symptoms, but suffer with debilitating brain fog.

Gut–endocrine connection

The microbiome helps in the modulation of various parts of the endocrine system, from hormone production to hormone secretion and metabolism. The gut also houses special cells called enteroendocrine cells that produce various hormones that are involved in appetite regulation, glucose metabolism and energy balance, released in response to nutrient intake, so you can see how our diet would impact how much of each are produced and secreted. The microbiome is also involved in the biotransformation of hormones, aiding the detoxification pathways of our body, for example some bacteria in the microbiome can metabolize and degrade hormones such as oestrogen as well as environmental toxins helping the body to eliminate them from the body.

Gut–thyroid connection

A healthy microbiome is crucial for a balanced thyroid. An enzyme produced by the gut is involved in the conversion of T4 to T3 (chapter 10) – 20% of this conversion happens in the gut.

Diversity

So how do we create and maintain diversity of the gut microbiome in order to keep us at optimal health?

Diversity of the microbiome is directly connected to diversity of our diet. Diversity in the diet is important because it ensures we are getting all the essential nutrients our body needs and it helps the beneficial bacteria to grow and thrive. This diversity has been linked to lower levels of inflammation, as well as reduced risk of bacterial imbalance in the gut, which can lead to digestive issues and other chronic health issues.

When we are looking to increase the diversity of nutrients in the diet we are looking to increase the range of phytonutrients that we eat. Phytonutrients are components of plants that are powerful defenders of health. Studies show that people who eat more plant foods have reduced risk of chronic illness, lower inflammation and better cognitive function. In Part 3 of this book you will find information and resources, such as the diversity checklist, to help you easily assess and increase your phytonutrient intake and diversity in the diet.

Intestinal permeability

You may have heard of something called 'leaky gut'; it's become a popular social media term to describe what we as practitioners call intestinal permeability. Healing intestinal permeability not only improves your gut health and ensures optimal functioning of the gut, it lowers inflammation in the body helping to prevent the development of chronic illness. Intestinal permeability is a known root cause of autoimmune diseases.

What is intestinal permeability?

Along our intestine wall we have cells called tight junctions. These tight junctions open and close to allow the smaller molecules from the food we eat to enter the bloodstream, such as vitamins and minerals. When we have intestinal permeability these tight junctions open wider (remember gluten helps this to occur), compromising our barrier function allowing larger molecules through into the bloodstream, including undigested food particles and toxins. These larger molecules would have continued on through our digestive tract and eventually been eliminated from the body. However, once this larger opening occurs they are now in our bloodstream and the only way for them to be eliminated from the body is for the immune system to be activated and come and mop them up. This may not sound like much of a problem, but there are a number of issues here. First, this creates inflammation in the body – think of when you cut your hand and how the area inflames as the immune system activates to keep pathogens out and heal your cut. Immune activation equals inflammation. Now where we need this immune response inflammation is a good thing, but if we are continually eating a diet that creates this inflammation, it becomes chronic inflammation and this is a bad thing. Second, this inflammation and immune activation can act as a trigger for autoimmune diseases and in those who already have autoimmunity it can create a flare up or a relapse, or aid in the development of another chronic disease.

Infections and toxicities

Dysbiosis

Dysbiosis isn't actually an infection, it is when you have an imbalance of the bacteria in your gut and this has been linked as a root cause to many chronic health conditions.

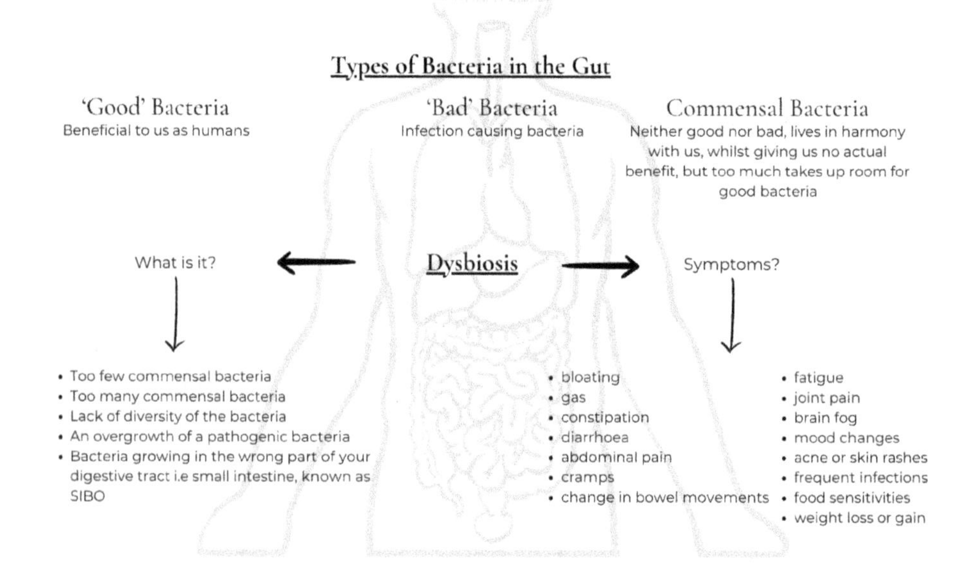

Parasites

Parasites are defined as an organism that lives off of the host in order to survive; they can even use nutrients from your diet to their own benefit. Within functional medicine, however, understanding parasites goes beyond the conventional notion of them as unwelcome invaders. We undertake a comprehensive view that their impact on gut health and their impact around the body are what is important. Parasites have the potential to disrupt the delicate balance of the gastrointestinal system, which can impact health within the gut and also our systemic health.

While eradicating parasites is part of the picture, it is not the whole story. We must address the underlying root causes that allow for them to grow and multiply. The integrity of our gut, the strength of our immune system and a diverse and well balanced microbiome are all root causes for parasites. As with everything else, addressing the root causes is key to restoring optimal health.

Parasites can be a cause of damage to the intestinal wall leading to intestinal permeability and can also disrupt the balance of the microbiome causing dysbiosis. Outside of gut health they can cause chronic inflammation and immune dysregulation. These in turn can lead to chronic conditions and autoimmune diseases, chronic fatigue, joint and skin disorders are common. Due to the parasites using nutrients from your diet, we can also see nutrient deficiencies. For some, parasites will not cause any particularly noticeable symptoms, you can be completely asymptomatic; for others, they can result in gastrointestinal symptoms and pain, weight loss, bloating, fatigue, skin itching, joint pain or aches, iron deficiency anaemia, allergic symptoms, headaches, cognitive dysfunction and brain fog.

Working on parasites includes targeted interventions to eradicate the parasite; we can assess what parasite you have within our comprehensive stool testing, alongside restoring gut barrier integrity, balancing the microbiome, supporting detoxification pathways and balancing the immune system.

Candida

Candida Albicans, or yeast, is a fungus that is naturally present in our system, however an overgrowth can cause significant issues. Candida can be a root cause but can also be a result of another imbalance. Disruption to the microbiome, antibiotic use, diets high in carbohydrates, sugars, processed foods and alcohol, chronic stress and inflammation can create an environment that is conducive to the growth of candida. If your immune system is impaired by poor sleep, nutrition, chronic stress or an underlying immune health condition, candida will be able to grow and take hold within your GI tract in a way it wouldn't if you had a healthy microbiome and gut function, strong immune system, low inflammation and adequate nutrient intake.

If candida is able to grow, it further takes up room in the gut microbiome pushing out the beneficial bacteria, causing dysbiosis and increasing the risk of intestinal permeability. It can also further impair the immune system creating low lying chronic inflammation and chronic activation of the immune system. Candida itself can cause gastrointestinal symptoms such as bloating, gas, abdominal pain, diarrhoea or constipation, while its impact on the immune system can mean symptoms such as fatigue, joint pain, headaches and recurrent infections occur. Candida has also been linked to conditions such as chronic fatigue syndrome, fibromyalgia, depression, anxiety and autoimmune diseases.

Viral

PEARL: Not only are we more bacteria than human, we are also more virus than human!

Yes while you may have heard of the microbiome, did you know there is also a human virome? Scientists now believe we are host to approximately 380 trillion viruses, either on or in our body – within our body many of these reside in the gut.

While some can cause illness, others are harmless and some are actually beneficial. We are complex superoganisms made up of more bacteria and viruses than human cells.

The virome, including both bacteriophages (viruses that infect bacteria) and eukaryotic viruses (viruses that infect eukaryotic cells), is increasingly acknowledged for its role in shaping gut health and overall wellbeing. The bacteriophages interact with the gut microbiome, shaping the composition and diversity of the gut environment and microbiome. The virome also influences our immune responses; bacteriophages can stimulate innate immune responses, which play roles in host defence against pathogens and maintenance of gut barrier integrity and can trigger adaptive immune responses, shaping the development of immune memory and immune tolerance. Viruses can also influence the production of SCFAs and the virome can also suffer imbalances that can lead to inflammation, immune dysregulation and host metabolism dysfunction.

Mould

Mould is a fungus which commonly grows in water-damaged or damp buildings, and releases mycotoxins and other by-products. Some such as Ochratoxin A and aflatoxins can damage the intestinal lining and compromise the barrier function. Ochratoxin A is also a persister mould. It binds to a protein called albumin in your blood. They can also promote inflammation and oxidative stress which can lead to cell death. Mould can also trigger immune responses in the gut, activating the innate immune system and stimulating release of pro-inflammatory cytokines and chemokines. Prolonged exposure can further activate the adaptive immune system. Mould has also been linked to development of dysbiosis, altered motility, secretion and nutrient absorption in the gut. Mould symptoms go beyond the gut as different moulds impact different parts of our body, immune function, cognitive symptoms such as brain fog and fatigue, inflammation, joint pain and stiffness are common symptoms.

Patient F came to see me with a wealth of symptoms such as brain fog, joint pain and fatigue, but had very few gastrointestinal symptoms, if any. We ran a batch of functional testing. Autoantibody testing came back with positives flagging up all over the place, showing me that something had triggered polyreactive autoimmunity. Polyreactive autoimmunity refers to a type of autoimmune response in which the immune system produces antibodies that can recognize and bind to multiple, diverse antigens or self-antigens. Polyreactive autoimmunity can contribute to the development and progression of autoimmune diseases through several mechanisms such as molecular mimicry and loss of tolerance. Now the challenge was to find what was causing this response all around the body. We ran viral testing, mycotoxin testing for mould, comprehensive stool test and adrenal testing. We found multiple infections and overgrowths. We had low lying viruses HHV-6 and cytomegalovirus, Ochratoxin A as a mould toxicity, dysbiosis Hafnia alvei as an imbalance bacteria and Citrobacter freundii complex as a dysbiotic bacteria, plus elevated cortisol on the adrenal stress test. None of these overgrowths were huge, but added up they were enough to cause intestinal permeability in the gut which caused inflammation and the polyreactive autoimmune response. This was causing the vast array of symptoms the patient was experiencing and if left could have led to an autoimmune disease. As it happened this patient went from struggling to walking down one flight of stairs to hiking. Gut health really can have a huge impact on our health.

The 5 Rs

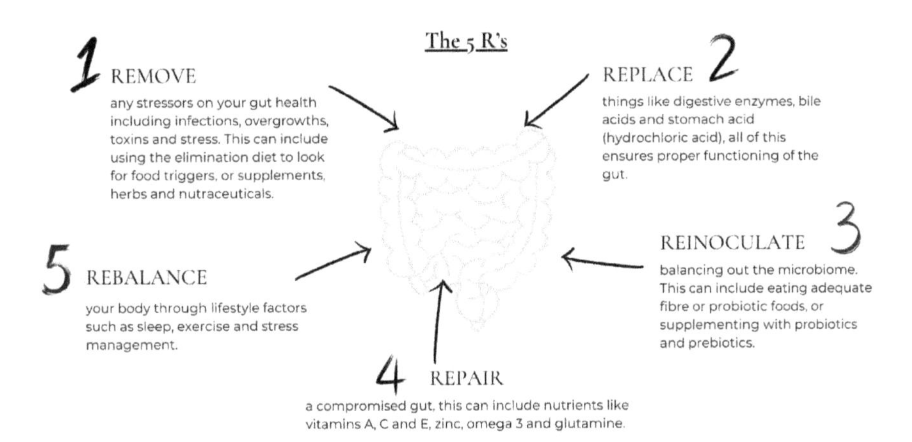

Food as a trigger

At times in our life food can be a trigger for symptoms or ill health. Sometimes we don't even realize that we are experiencing food triggers until they are removed from the diet. We are aware of food allergies where you get a more severe response to eating a food, but we also have food intolerances and sensitivities. These are usually a result of an imbalance in the gut. Food as a trigger can lead to low level inflammation, damaging the gut lining and causing intestinal permeability, as well as causing dysbiosis.

The elimination diet is a functional medicine tool that is a gold standard way of finding food triggers personalized to you. We want to remove as few foods from your diet as possible, as we have seen diversity in the diet is important to optimal health. We only want to remove the foods that impact you personally and then heal the gut to get as many foods back in as possible.

The elimination diet has to be done in a very specific way in order to see true results. If you want to undertake an elimination diet you can go to the online book resources section where you will find a six-week elimination programme showing you exactly how to run the diet with recipes and meal plans. The elimination diet is a short-term food plan and should not be seen as a long-term lifestyle programme – it is a part of your gut healing journey.

We can also run food intolerance testing, there are different qualities of food intolerance testing and many of the at home kits you can buy are extremely limited. At Goode Health we use a specific type of food sensitivity testing that we can discuss with you.

The gut nutrients

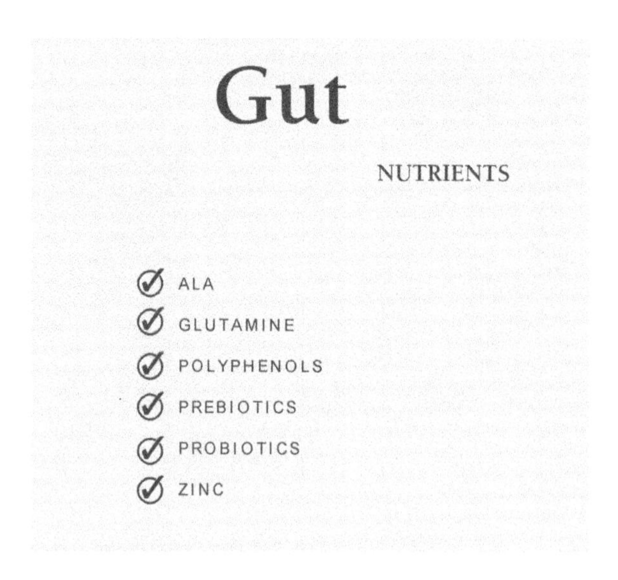

12

Hormone health

Overview of the major hormones and their functions in the body

Hormones are chemical messengers produced by various glands throughout the body, playing pivotal roles in regulating essential functions, maintaining homeostasis and supporting overall health. They travel around our body in the bloodstream, or in the fluid surrounding our cells; neurotransmitters travel via the central nervous system.

PEARL: Hormones are your body's messengers; the messages they deliver change the way your cells act.

Understanding the major hormones and their functions is crucial in comprehending the intricate interplay within the endocrine system. From controlling metabolism to influencing mood and reproduction, hormones orchestrate a symphony of bodily processes. The endocrine system comprises glands that secrete hormones directly into the bloodstream, influencing target organs and tissues. Key players in this system include the pituitary gland, thyroid gland (chapter 10), adrenal glands (chapter 9), pancreas, ovaries, and testes. Each gland produces specific hormones that regulate diverse physiological functions, ensuring coordination and balance throughout the body.

Insulin

Insulin is produced by the pancreas and regulates blood glucose levels (chapter 13) by facilitating the uptake of glucose into cells for energy production. Imbalances in insulin levels can lead to diabetes mellitus and metabolic disorders.

Growth hormone (GH)

Secreted by the pituitary gland, GH stimulates growth, cell regeneration and repair. It plays a crucial role in childhood growth and continues to regulate metabolism and tissue maintenance throughout life.

Oestrogen and progesterone

These are sex hormones and are predominantly produced by the ovaries (and in smaller amounts by the adrenal glands); they regulate menstrual cycles, reproductive function, bone health and cardiovascular health in women. They also influence mood, cognition and skin health.

Testosterone

Another of the sex hormones, primarily produced in the testes (and in smaller amounts in the ovaries and adrenal glands), testosterone is vital for male reproductive function, muscle mass, bone density, libido and mood regulation. It also contributes to female health.

Melatonin

Synthesized by the pineal gland, melatonin regulates the sleep–wake cycle (chapter 9) and supports immune function. It also exhibits antioxidant properties, protecting against oxidative stress (chapter 6) and promoting overall health.

Serotonin and dopamine

Neurotransmitters produced in the brain (chapter 8) play key roles in mood regulation, cognition, reward pathways and motor control. Imbalances in these neurotransmitters are associated with various mental health disorders.

Importance of hormonal balance for overall health and longevity

Maintaining hormonal balance is essential for overall health and longevity. Imbalances or dysregulation of hormones can lead to a myriad of health issues, including metabolic disorders, reproductive dysfunction, mood disorders and impaired cognitive function. Factors such as genetics, lifestyle, diet, stress and environmental toxins can influence hormone levels and disrupt the delicate equilibrium within the endocrine system. The endocrine system is heavily interconnected – there will be more than one hormone involved in homeostasis and balance of a particular area.

Patient D came to see me with various symptoms such as loss of muscle mass, sexual dysfunction, poor libido and low energy. Rather than just focusing solely on the sex hormones, we looked for the underlying imbalances and root causes. We found adrenal imbalance and the 'cortisol steal' in play (chapter 9). His immune function was also affected and we found autoantibodies in his blood work putting him at greater risk of certain autoimmune conditions as well as markers indicating intestinal permeability, which we have already seen is connected to the immune function (chapter 11). We found loss of chemical tolerance where his immune function was impaired and importantly impaired detoxification with high levels of inflammation and oxidative stress in the body. He had a risk marker for autoimmune liver conditions and his liver was struggling to cope with the chemical load in the body that likely built up due to impaired immune function. We will come on to the importance of the liver and detoxification in the text box on page 100, but the important thing to see here is that all of this led to dysfunction and imbalance of his sex hormones, showing how interconnected the hormones and endocrine system is around the body.

Hormones play a significant role in modulating immune function (chapter 7) and inflammatory responses. For instance, cortisol has immunosuppressive effects, while hormones like oestrogen and testosterone can influence immune cell activity and cytokine production. Dysregulation of these hormones can disrupt immune function, leading to increased susceptibility to infections, autoimmune diseases and chronic inflammation.

Hormone balance is intricately linked to the ageing process and longevity. As individuals age, hormone levels naturally decline, contributing to age-related changes in metabolism, muscle mass, bone density and cognitive function. By maintaining hormonal equilibrium we can enhance vitality, resilience and overall quality of life as we age.

Understanding the endocrine glands and their roles

The endocrine system consists of several glands scattered throughout the body, each producing specific hormones that exert diverse effects on target cells and tissues. These glands can be categorized into primary and secondary endocrine glands based on their anatomical location and primary function.

Primary endocrine glands include the hypothalamus, located in the brain, which serves as the master regulator of the endocrine system. It produces releasing and inhibiting hormones that control the secretion of hormones from the pituitary gland, thus exerting indirect control over various endocrine functions. The pituitary gland is often referred to as the 'master gland'; the pituitary is situated at the base of the brain and consists of two lobes: the anterior pituitary (adenohypophysis) and the posterior pituitary (neurohypophysis). It secretes a myriad of hormones that regulate growth, metabolism, reproduction, stress response and other essential functions.

Secondary endocrine glands include the thyroid gland (chapter 10) and the parathyroid glands (which are four small glands located on the posterior surface of the thyroid gland). The parathyroid glands produce parathyroid hormone (PTH) that regulates calcium and phosphate levels in the blood and bone. The adrenal glands (chapter 9) are where the adrenal cortex produces steroid hormones such as cortisol, aldosterone and sex hormones, and the adrenal medulla secretes catecholamines (adrenaline (epinephrine) and noradrenaline (norepinephrine)). The pancreas is an organ with both endocrine and exocrine functions; the pancreas produces hormones such as insulin and glucagon. Islets of Langerhans within the pancreas contain alpha and beta cells responsible for hormone secretion. The gonads, ovaries and testes produce oestrogen, progesterone and testosterone, respectively.

Detoxification pathways

The liver

The liver is not an endocrine organ, but no discussion around hormones would be complete without considering the liver and detoxification. The liver is one of the body's vital organs and is involved in metabolism, digestion, detoxification and hormonal regulation. Most of you probably know the liver for processing nutrients and filtering out toxins, but it also has its role to play in the endocrine system producing and metabolizing hormones.

The liver is a large, reddish-brown organ located in the upper right quadrant of the abdomen, just beneath the diaphragm. It is essentially a giant filter for all blood coming from the intestines. The liver metabolizes nutrients, converting proteins, carbohydrates and fats into energy or storing them for future use; the liver also acts as storage for vitamins (like A, D, E, K), minerals (iron, copper) and glycogen. It synthesizes glucose, processes amino acids (your protein building blocks) and regulates lipid (fat) metabolism. Hepatocytes (cells in the liver) produce bile, which is stored in the gallbladder, a digestive fluid that helps in the breakdown and absorption of fats in the small intestine. The liver plays a central role in detoxification, removing harmful substances such as drugs, alcohol, metabolic waste products and environmental toxins from the blood stream (see case study box on page 99). The liver can metabolize and eliminate these toxins with the help of enzymes to make them more water soluble for excretion. The liver also metabolizes and clears hormones from the bloodstream, maintaining hormonal balance and endocrine regulation.

There are four main ways we can detoxify substances from our body – through our SUBS:

1. Stool
2. Urine
3. Breath
4. Sweat.

For the stool and urine we are looking to phase I and phase II detoxification in the liver. The final stage of elimination is via the gut through the stool or via the kidneys through urine.

Phase I Detoxification

In phase I detoxification, hepatocytes use enzymes such as cytochrome P450 to metabolize metabolites made in the body and toxins into intermediate metabolites. These reactions make the toxins water-soluble for further processing, however it also makes them more reactive, increasing the risk of harm from these substances.

Phase II Detoxification

In phase II detoxification, the intermediate metabolites produced in phase I are made less reactive and tags them for excretion. This conjugation process renders the toxins less toxic and more readily excreted, via bile into stool through the intestines, or via urine. If one or both of these phases are out of balance, hormones such as oestrogen are not eliminated and they can recirculate resulting in oestrogen dominance.

We can support liver function by consuming nutrients the liver requires for both phase I and phase II of detoxification and reducing toxin load (chapter 18). Different nutrients are required by the liver as cofactors for the different stages.

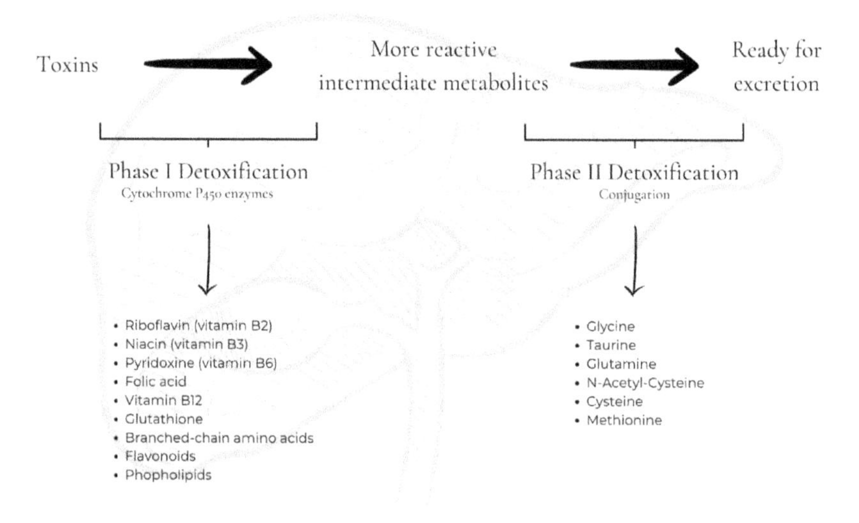

Research has shown that cruciferous vegetables, soy, berries, garlic and spices can support detoxification of polychlorinated biphenyls (PCBs). Phytonutrients such as curcumin, quercetin and resveratrol have been found to aid in neutralizing environmental toxins that are endocrine disruptors (chapter 18). Phytonutrients, especially those in cruciferous vegetables, regulate liver enzymes and polyphenols also support biotransformation, especially of oestrogen.

We can also breathe out toxic substances such as carbon dioxide and we have to remember that breathing in is a potential source of inhaling toxins. Air pollution and things like moulds (mycotoxins) can cause lung inflammation, impaired immune function, autoimmune disease, respiratory disease and more. Sweating is another way in which we can support detoxification – some substances such as persistent organic pollutants (POPs) are actually better excreted via sweating. Heavy metals are also well excreted in sweat more than in urine, which is why I often recommend saunas to people I work with (chapters 17 and 18).

Methylation

In order to optimize our health on a cellular level we can't ignore methylation. I could have talked about methylation in many different parts of this book because the reality is that it is involved in energy production, immune function, neurotransmitters in the brain, DNA synthesis, the cardiovascular system, the reproductive system, hormone balance, detoxification and more.

Methylation is a biochemical and epigenetic process that involves the transfer of a methyl group (CH3) from one molecule to another. Methylation is catalyzed by enzymes known as methyltransferases, which transfer methyl groups from molecules such as S-adenosylmethionine (SAMe) to specific substrates, thereby modifying their structure and function. The compounds that require methylation, such as proteins, enzymes and hormones, need it in order to function optimally.

PEARL: Methylation happens approximately 1 billion times a second, affecting nearly every process in the body.

Methylation supports the natural detoxification pathways, aiding the biotransformation and elimination of hormones and toxins. It also helps production of glutathione, our master antioxidant (chapters 6 and 7), which aids in the removal of toxins by binding to them.

Suboptimal methylation can lead to low energy, impaired detoxification, impaired immune function and more. While genetics plays a part, we can optimize this process via dietary and lifestyle factors. Let's take a look at some of the nutrients which can impact methylation.

Folate (B9)

Folate serves as a methyl donor in methylation reactions. Folate is converted into its active form, 5-methyltetrahydrofolate (5-MTHF), which donates a methyl group for methylation reactions, including the conversion of homocysteine to methionine. Methionine, in turn, serves as the precursor for S-adenosylmethionine (SAMe), the primary methyl donor in numerous methylation reactions throughout the body.

Riboflavin (Vitamin B2)

Riboflavin is a cofactor for the enzyme methylenetetrahydrofolate reductase (MTHFR), which plays a crucial role in folate metabolism. MTHFR catalyzes the conversion of 5,10-methylenetetrahydrofolate to 5-MTHF, the active form of folate involved in methylation reactions. Riboflavin deficiency can impair MTHFR activity, leading to decreased availability of 5-MTHF for methylation processes.

Cobalamin (Vitamin B12)

Vitamin B12 is involved in the conversion of homocysteine to methionine, a critical step in methylation reactions. Vitamin B12 acts as a cofactor for the enzyme methionine synthase, which transfers a methyl group from 5-MTHF to homocysteine, generating methionine. Methionine serves as the precursor for SAMe, the primary methyl donor in methylation reactions.

Methionine

Methionine is an essential amino acid that serves as the precursor for SAMe, the primary methyl donor in methylation reactions. SAMe donates its methyl group to various substrates, including DNA, proteins, neurotransmitters and lipids, facilitating methylation reactions throughout the body.

Pyridoxine (Vitamin B6)

Vitamin B6 is involved in various aspects of one-carbon metabolism and methylation pathways. Pyridoxal phosphate (PLP), the active form of vitamin B6, acts as a cofactor for enzymes involved in amino acid metabolism, generating cysteine, a precursor for glutathione synthesis.

Choline

Choline is a vitamin-like nutrient that serves as a precursor for betaine, a methyl donor involved in methylation reactions. Betaine is synthesized from choline via a pathway known as the choline oxidation pathway. Betaine donates a methyl group to homocysteine, converting it to methionine. Betaine serves as an alternative methyl donor when folate or vitamin B12 levels are insufficient.

Maintaining balance and optimizing sex hormones

As we age, hormonal imbalances can occur; I like to call it the hormonal highway as it's up and down and all over the place, especially for woman. Changes in hormone levels can have a significant impact on healthspan as your sex hormones, including oestrogen, progesterone and testosterone, play crucial roles in regulating various physiological processes essential for wellbeing. Quality of life is one of the biggest things I hear my patients say in clinic about hormonal imbalances; they feel the symptoms significantly impact their overall health and ability to be their best selves. As we age, changes in sex hormone levels can contribute to a decline in physiological function, cognitive function, mood imbalances and increase the risk of age-related diseases. For example, declining oestrogen levels during menopause can lead to symptoms such as hot flashes, vaginal dryness, mood swings, weight gain and increased risk of osteoporosis and cardiovascular disease. Decreased progesterone levels during perimenopause and menopause can contribute to irregular menstrual cycles, sleep disturbances, anxiety and mood swings. Declining testosterone levels with age can result in symptoms such as reduced libido, decreased muscle mass, fatigue, depression and cognitive decline.

Hormones, digestion, nutrient absorption and the gut health connection

The largest organ to produce hormones is the gut and it produces a huge quantity of them. The digestive system relies on a variety of different hormones to regulate various processes, including digestion, nutrient absorption, appetite control and gut motility.

Ghrelin is often referred to as the 'hunger hormone', it stimulates appetite, while leptin, known as the 'satiety hormone', signals fullness and regulates energy balance. Insulin and glucagon play crucial roles in glucose metabolism and blood sugar regulation, influencing nutrient uptake and utilization. Glucagon-like peptide-1 (GLP-1), peptide YY (PYY), cholecystokinin (CCK) and gastric inhibitory peptide (GIP) are released in response to food intake, stimulating insulin secretion, slowing gastric emptying, and promoting satiety.

Hormonal influences on cognitive function and optimal health

Hormones are intricately linked to cognitive function and optimal health. Many hormones have profound effects on the brain and are linked to cognitive ageing.

1. Hormones such as oestrogen, progesterone, testosterone, cortisol, insulin, thyroid hormones and growth hormones are involved in regulating neuronal function, synaptic plasticity, neurogenesis and cognitive processes such as memory, attention and executive function.
2. Mostly in women oestrogen and progesterone support cognitive function, mood regulation and are neuroprotective.
3. In men testosterone influences spatial cognition, memory and mood regulation, impacting cognitive ageing and neurodegenerative diseases.
4. Cortisol, the main stress hormone, can have both beneficial and detrimental impacts on cognitive function. Cortisol levels have been implicated in cognitive impairment, hippocampal atrophy and accelerated brain ageing.
5. Insulin resistance is associated with cognitive decline, Alzheimer's disease and vascular dementia, highlighting the importance of insulin sensitivity for brain health.
6. Thyroid hormones play essential roles in neuronal development, myelination and synaptic transmission, with hypothyroidism and hyperthyroidism both impacting cognitive function and mood regulation.
7. Growth hormone and insulin-like growth factor 1 (IGF-1) support neuronal survival, synaptic plasticity and cognitive function, with implications for cognitive ageing and neuroprotection.

The hormone nutrients

Hormone

NUTRIENTS

- ALA
- B VITAMINS
- CHOLINE
- GLUTATHIONE
- MAGNESIUM
- OMEGA-3 FATTY ACIDS
- VITAMIN D
- ZINC

13

Cardiometabolic health

Importance of maintaining cardiometabolic balance for optimal health and longevity

Cardiometabolic health encompasses the balanced interplay between cardiovascular health and metabolic processes. Disruptions in this balance can predispose individuals to a spectrum of conditions. In the contemporary world we live in, characterized by sedentary behaviours, poor dietary habits, chronic stress and environmental toxins, maintaining cardiometabolic balance has become increasingly challenging, which has all fuelled an epidemic of obesity and metabolic syndrome. The good news is much of this comes down to factors that can be moderated, bringing the risk down and reversing metabolic disorders.

PEARL: While chronic illnesses like autoimmune diseases are the leading cause of poor quality of life, cardiovascular diseases are the leading cause of deaths globally.

Cardiovascular health is associated with a lower risk of premature mortality as well as a longer healthspan. Healthier nutrition and dietary habits are associated with a cardiovascular disease (CVD) free life expectancy between the ages of 55 and 85 years. Research in 2023 showed that a mix of a preventative approach with lifestyle and dietary changes and targeted interventions for current cardiometabolic disorders (CMDs) could delay cognitive decline.

PEARL: 17.9 million people died from cardiovascular diseases in 2019 (globally), 85% of these were heart attack or stroke.

Worryingly CMDs are growing in those aged 25–44 years. It is thought the modern western diet, fast food and lifestyle changes, such as more sedentary jobs and increased chronic stress, play a role in this.

PEARL: Globally, in 2019, of 17 million premature deaths (under 70) due to noncommunicable diseases, 37% were caused by cardiovascular disease.

Metabolic syndrome is a cluster of risk factors for CVDs. Metabolic disorders are on the rapid rise and include things like obesity, insulin resistance, impaired glucose metabolism, high cholesterol and high blood pressure. Studies have shown a causal link between these metabolic markers and increased risk of CVDs like coronary heart disease, myocardial infarction, heart failure, hypertension and stroke.

The conditions

Cardiometabolic disorders (CMDs)

CMDs encompass a spectrum of conditions that involve both cardiovascular and metabolic abnormalities. These disorders often coexist and share common risk factors, including obesity, hypertension, dyslipidemia, insulin resistance and inflammation. Examples of CMDs include metabolic syndrome, type 2 diabetes and non-alcoholic fatty liver disease. They increase the risk factor of developing a cardiovascular disease.

Cardiovascular disease (CVD)

CVD refers to a group of conditions that affect the heart and blood vessels, including coronary artery disease, heart failure, stroke, peripheral artery disease and other conditions. These diseases typically result from atherosclerosis, a process in which fatty deposits build up in the arteries, leading to narrowing or blockage and reducing blood flow to vital organs.

Metabolic syndrome

Metabolic syndrome is a cluster of metabolic abnormalities that increase the risk of developing CVDs and type 2 diabetes. The defining criteria for metabolic syndrome typically include abdominal obesity, elevated blood pressure, dyslipidemia, impaired glucose metabolism and insulin resistance.

Dyslipidemia

Dyslipidemia refers to an abnormality in the levels of lipids (fats) in the blood, including cholesterol and triglycerides. It encompasses elevated levels of total cholesterol, low-density lipoprotein cholesterol (LDL or 'bad' cholesterol) and triglycerides, or reduced levels of high-density lipoprotein cholesterol (HDL or 'good' cholesterol). Dyslipidemia is a significant risk factor for CVDs and often coexists with other CMDs such as obesity, insulin resistance and metabolic syndrome.

Impaired glucose metabolism

Impaired glucose metabolism refers to abnormalities in the body's ability to regulate blood sugar levels effectively. This can manifest as impaired fasting glucose (elevated fasting blood sugar levels) or impaired glucose tolerance (elevated blood sugar levels after consuming a meal). Impaired glucose metabolism is considered a precursor to type 2 diabetes and is often present in those with metabolic syndrome or insulin resistance.

Insulin resistance

Insulin resistance is a condition in which cells in the body become less responsive to the effects of insulin, a hormone produced by the pancreas that helps regulate blood sugar levels. As a result, higher levels of insulin are required to maintain normal blood glucose levels. Insulin resistance is a key feature of metabolic syndrome and type 2 diabetes.

Pre-diabetes

Pre-diabetes is a condition in which blood sugar levels are higher than normal, but not yet high enough to meet the diagnostic criteria for diabetes. Pre-diabetes is often diagnosed based on specific blood glucose measurements. Individuals with pre-diabetes have an increased risk of developing type 2 diabetes, CVDs and other complications.

Diabetes

Diabetes is a chronic condition characterized by elevated blood sugar levels resulting from either not producing insulin (type 1 diabetes) or insufficient insulin production, ineffective use of insulin by the body or a combination of both (type 2 diabetes).

PEARL: Losing 7% of excess weight can reduce onset of type 2 Diabetes Mellitus by 58%.

Blood sugar balance

Blood sugar, also known as blood glucose, refers to the concentration of glucose, which is a type of sugar, present in the bloodstream. I'm guessing you probably already know something about blood sugar levels, most of us do these days. The 2021 figures show us that almost a third of Americans aged 18 or over have pre-diabetes – that's over 90 million people and another 38 million have diabetes. In 2023 the number of people with diabetes in the UK went over 5 million for the first time.

PEARL: Each week, diabetes leads to over 770 strokes, 590 heart attacks and 2300 cases of heart failure.

The process is quite simple, the more sugar you eat, the more likely your blood sugar is to be affected. When you eat carbohydrates, particularly if you eat them without fat or fibre (and this is why a balanced plate is so important as you will see in Part 3 of this book), the food is converted into glucose and the release of insulin is triggered. The insulin takes the sugar out of your blood and sends it where it needs go, for example to your muscles, or to your liver, to be stored. Too much sugar and your insulin can struggle to keep up and may spike and stay high for too long.

A word on continuous glucose monitors

Now I want to add a note here, because recently there's been lots of noise around continuous glucose monitors (CGMs) and not having blood sugar spikes. We have to be very careful around this messaging. When we eat, our blood sugar levels rise. Now for people with diabetes these CGMs monitor your blood sugar 24/7, which can be useful and even life changing, but they are increasingly used in the nutrition world in healthy people. CGMs have gone from being a medical tool to a commercialized piece of tech. On the one hand it's good that awareness is being raised around blood sugar, as we've seen it's a huge problem. My concerns are these. People are using these and looking at standardized levels for blood sugar ranges for pre-diabetes and think blood sugar should never go over this level. These levels weren't designed to be used in this way as a 24/7 marker. There is also little to no scientific evidence on the use of CGMs in healthy individuals, i.e. those without diabetes (there is in people with diabetes). Continuous monitoring can cause health anxiety; if levels go up at any time people think they have diabetes. Another concern I have is that constantly following your blood sugar may lead to people swapping out some healthy foods, such as fruit or even carbs, for high fat foods that spike the blood sugar less, but actually may put them at more risk of developing diabetes. And finally I think this continuous monitoring in a healthy person can lead to disordered thoughts and approaches to eating, putting people at risk of things like eating disorders. So CGMs have their use and can be hugely beneficial to people with diabetes, but we have to be careful that we are using them appropriately. It's important to remember that our blood sugar levels fluctuate and this is normal. What we don't want is constant highs or a permanent rollercoaster picture.

Symptoms of blood sugar imbalance

When blood sugar levels are persistently elevated, it can lead to a condition known as hyperglycaemia – you may experience symptoms such as increased thirst and frequent urination due to the excess glucose in the bloodstream causing the kidneys to excrete more water, which can cause dehydration. It can also impact the fluid balance in the eye leading to blurred vision. You may also experience fatigue and weakness if cells are unable to effectively utilize the glucose for energy production. You can experience hunger which can lead to you reaching for a snack, often a sugary snack, further elevating blood sugar levels. It can also impact the immune function leaving you prone to infections and with impaired wound healing.

When we get to insulin resistance, where your cells are now less responsive to the insulin, your cells may be deprived of glucose despite elevated blood sugar levels and you can experience increased hunger and cravings for carbohydrates. Your cells may also not effectively use glucose for energy production leading to fatigue. Insulin resistance can also lead to weight gain as it promotes storage of fat, especially around your middle.

Symptoms of diabetes can see all of the previously mentioned symptoms at increased levels with more fatigue, thirst, urination, weakness and blurred vision. It can also lead to further impaired immune function.

Cholesterol and your lipid profile

Cholesterol plays a crucial role in cardiometabolic health. It is a structural component of cell membranes and a precursor for the synthesis of steroid hormones, bile acids and vitamin D. While cholesterol is essential for various physiological processes, abnormal levels and distribution of cholesterol in the bloodstream can contribute to the development of cardiometabolic disorders, particularly atherosclerosis and cardiovascular disease.

Elevated levels of LDL cholesterol promote the deposition of cholesterol-rich plaques in arterial walls, leading to the narrowing and hardening of blood vessels and increasing the risk of coronary artery disease, heart attack and stroke. Conversely, HDL cholesterol plays a protective role by transporting excess cholesterol from peripheral tissues back to the liver for excretion, thereby reducing the risk of atherosclerosis and cardiovascular events. Triglycerides, another type of lipid found in the bloodstream, can also influence cardiovascular risk, particularly when levels are elevated in conjunction with low HDL cholesterol.

Patient K came to see me with no diagnoses, but she had elevated blood sugar having been told she was prediabetic, an elevated total cholesterol with high levels of LDL and low levels of HDL cholesterol. She was experiencing fatigue and some brain fog, but had put this down to being busy and getting that bit older. This is a common response I see in people, but particularly in women. After we worked together, patient K said to me that she had forgotten what it felt like to be well because for so long she had accepted niggling symptoms as the norm and something she just had to put up with. When we looked into her history we found little of significance. Patient K had been well through most of her life, with the main pressures on her body being two pregnancies, both of which were straightforward and quickly and easily recovered from. What we did find was a pattern of poor sleep since having the children. Patient K had never really regained a good circadian rhythm, despite her children now being older (read more on the impact of sleep on page 161). We also found some stress: patient K worked, had two children and an ailing parent. Nothing out of the ordinary, just normal life stressors, but mixed with poor sleep we had two of the main lifestyle factors for cardiometabolic health impaired. Patient K was still exercising, but was finding that recovery was a little harder than it used to be. Her immune function also wasn't as good as it had been; she had been the person who never really picked anything up, but as we looked at her timeline she realized that actually she had had more in the way of viruses of late and probably didn't have the robust immunity she used to. Again this was not a major factor for her and if those markers in her yearly bloods hadn't been picked up she probably would have accepted the lowered immunity without even realizing.

When we looked at her cardiovascular risk she was actually relatively high – she had a grandparent who had suffered a heart attack and one who had suffered a stroke, which she hadn't really given much thought to in relation to her own health. It was only when we pieced this all together – the blood sugar results, the cholesterol profile, the familial history, the long-term impaired sleep pattern, the stress and her adrenal results showing stage 2 HPA axis dysfunction, the lowered immunity, plus her fatigue and food cravings – that she started to see the impact all of this together was having on her health. We worked to bring her blood sugar back to balance and lower her cholesterol through nutritional, lifestyle and supplement changes. We implemented a sleep routine and stress management techniques and added in some adaptogenic supplements for her adrenals (more on this in Part 3).

Within eight weeks, patient K was feeling much improved with better energy and sleep and better cognitive function, that she hadn't even really realized was being impaired. She felt her memory, performance and decision-making was better, feeling generally 'sharper'. We later retested her blood sugar and cholesterol; she was no longer prediabetic but in normal range, her total cholesterol was down and her ratio had started to improve. Her CVD risk had improved dramatically.

Cardiometabolic health around the body

Metabolism is tightly regulated by a complex interplay of hormones (chapter 12). Metabolism is also involved in production of ATP (chapter 6). The gut microbiome (chapter 11) impacts metabolism through its intricate interactions with host physiology, dietary components and microbial metabolites. Production of short-chain fatty acids (SCFAs) serve as energy sources for intestinal epithelial cells and contribute to metabolism. Additionally, the gut microbiome influences the metabolism of dietary nutrients and bile acids and dysbiosis (chapter 11) has been implicated in the pathogenesis of metabolic disorders. Stress contributes to the development of metabolic disorders as chronically elevated cortisol levels can disrupt normal metabolic processes, leading to insulin resistance, abdominal obesity, dyslipidemia and hypertension. Additionally, stress-induced alterations in appetite and food intake, coupled with increased cravings for high-calorie, high-fat foods, contribute to weight gain and metabolic dysfunction. Chronic stress also promotes the release of inflammatory cytokines and free radicals, exacerbating oxidative stress and inflammation, which are implicated in the pathogenesis of metabolic disorders and cardiovascular disease. Inflammation can impact atherosclerotic plaque formation and impact signalling pathways, while CMDs in turn can exacerbate systemic inflammation. Emerging evidence suggests that metabolic abnormalities contribute to cognitive decline and increase the risk of neurodegenerative disorders such as Alzheimer's disease and vascular dementia. Furthermore, vascular risk factors associated with cardiometabolic dysfunction compromise cerebral blood flow and impair neuronal connectivity, further exacerbating cognitive impairment.

Nutrition and lifestyle factors for metabolic disorders

Nutrition is a cornerstone of cardiometabolic health and can even help reverse CMDs. Dietary choices influence factors such as blood sugar regulation, lipid metabolism and blood pressure. The MitoImmune Programme in chapter 14 of this book is designed to support cardiovascular health through a balanced diet rich in fruits, vegetables, whole grains, lean proteins and healthy fats. This can help support optimal metabolic function and reduce the risk of obesity, insulin resistance, dyslipidemia and hypertension.

Adequate sleep (chapter 16) is essential for cardiometabolic health, as it influences various physiological processes, including glucose metabolism, appetite regulation and inflammation. Chronic sleep deprivation or poor sleep quality has been associated with disruptions in hormonal regulation, increased appetite and cravings for high-calorie foods, impaired glucose tolerance and elevated levels of inflammatory markers, contributing to the development of metabolic disorders and increasing the risk of CVD and type 2 diabetes.

Chronic stress (chapter 16) is a significant contributor to cardiometabolic dysfunction, as it activates the hypothalamic-pituitary-adrenal (HPA) axis and the sympathetic nervous system (chapter 9). Prolonged exposure to stress can disrupt metabolic homeostasis, promote insulin resistance, dyslipidemia, hypertension and abdominal obesity, and contribute to the development of CVD and type 2 diabetes.

Regular movement of your body is a cornerstone of cardiometabolic health benefiting many physiological systems, including cardiovascular, metabolic and immune function. Physical activity improves insulin sensitivity, enhances lipid metabolism, reduces blood pressure and promotes weight loss and weight maintenance. Additionally, exercise has an anti-inflammatory effect and can improve endothelial function, vascular health and mood.

The cardiometabolic nutrients

Cardiometabolic

NUTRIENTS

- ALA
- B VITAMINS
- CHOLINE
- CoQ10
- FLAVANOIDS
- GLUTATHIONE
- MAGNESIUM
- OMEGA-3 FATTY ACIDS
- POLYPHENOLS
- PQQ

PART 3

The MitoImmune way – your path to optimal health

14

The MitoImmune programme

So you plan to live well for longer. I'm assuming that statement is true if you have got this far. I am too. Who wouldn't want to? The good news, as we have seen, is that emerging research and science is giving us more and more answers on how to do this. You can actively start putting steps in place today that will support you on this journey and the steps don't need to be extreme or complicated.

PEARL: Simple changes equal BIG change.

The MitoImmune Programme is here to support you, to improve your health in the present and protect your health for the future.

PEARL: So many things in our life seem important. Until you become ill. Once illness hits you realize that the only thing that ever really mattered was your health.

You may have many goals for your life. Maybe you want to travel the world, or your focus is on being there to see your children grow up and provide them with the life you want to build, it might be that your goals are success related, you want a promotion, you want to build your own business and legacy, or you just want to be at the top of whatever game you are in, maybe you want to climb Kilimanjaro, or instead you want to go and lie on a beach in the Caribbean. The fact is, whatever you want to do with your life, however simple or extravagant your ambitions are, without your health you can't do any of it. The simplest of tasks can become impossible, never mind all the things you actually want to do. Trust me I've been there (and back).

The MitoImmune Programme is your key to optimizing your health, whether this comes from a place of wellness to reach optimal performance, or a place of chronic illness and wanting to rebalance your body; this programme is designed to optimize, balance and prevent. All too often I hear people saying 'oh it's just age' or 'it's not that bad' or 'it's normal for me'. It's time to put an end to this. It's time to realize that we all deserve to live well.

When I look back on my own journey I see now what I couldn't see then. I see underlying root causes such as viral load, I see burnout and stress, I see the onset of autoimmunity. Now I can see all the ups and downs of my own wellness journey and I know why – knowing ***why*** is power. I can look back and see a version of me who accepted a new normal, a normal that I shouldn't have accepted. I can see a version of me that just kept pushing through with no answers and no route to recovery. I can see a version of me that sacrificed health, because I didn't know what else to do; I was at a loss, getting no answers and no medical support. Then I found functional medicine and the new me started to emerge, or actually the old me, pre my viral trigger, reappeared. Now in my 30s I see what I ought to have always known. ***Health matters more than anything and we should always trust our instincts, listen to our bodies and never accept anything as 'normal'.***

PEARL: The key lies in root causes, in asking why, in optimizing your health and finding balance.

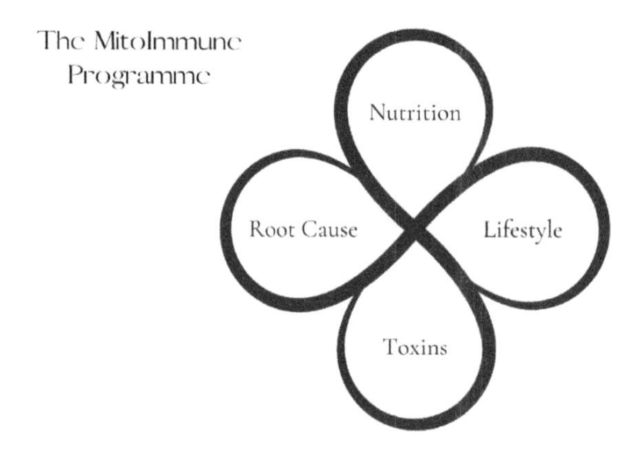

The MitoImmune Programme consists of four parts: Nutrition (chapters 14 and 15), Lifestyle (chapter 16), Toxins (chapters 17 and 18) and Root Cause (chapter 19).

How to implement The MitoImmune Programme:

1. The MitoImmune Nutrition Plan (this chapter) is designed to cover all eight of the pillars of health we have looked at. This anti-inflammatory programme filled with antioxidants will support each area of your health and is the foundation of your journey, whether you are optimizing your health or balancing a chronic condition. With simple steps you can go all in on the whole plan, or introduce one step at a time. This plan is helping you to:

- Support energy levels
- Support cognitive function
- Reduce inflammation

- Reduce oxidative stress
- Balance your immune function
- Support your cardiac and metabolic function
- Balance the gut microbiome
- Balance your hormones
- Reduce toxic exposure and detoxification pathways.

2. The MitoImmune Health Assessment – if you haven't already done so, jump to the online book resources or my website and take the MitoImmune Health Assessment (www.nicolegoodehealth.com/mitoimmune-health-assessment); learn about your pillars and where you may need to focus some attention. Look at the report you receive – the lowest scoring areas are where you should start your deeper dive work.
3. Specific Nutrients – if you find a pillar of your health comes back low you may want to focus in on this area. At the end of each chapter you had a checklist for which specific nutrients are supportive. Look at your assessment results, take your lowest scoring area and have a look at the end of that chapter at which nutrients are going to further boost support for you. Then in chapter 15, learn more about those nutrients, which foods they are in and how to incorporate them into your diet.
4. Lifestyle: the 4 S's – you can start slow with the nutrition plan and take it step by step, or you can go all in and work on the nutrition and the lifestyle factors at the same time. Either introduce this here, or start on this from the beginning of your journey. You will find steps to support sleep, stress, strength and social aspects of your life – the 4 Lifestyle S's (chapter 16).
5. Reduce Toxin Exposure – next we look at how you can start to reduce the toxin load in your body. The quality of our food (chapter 17) and environmental toxins (chapter 18) will help you to lower the burden.
6. Functional Testing – for those of you who are ready to deep dive into your health, find underlying imbalances and get to the root cause, this is for you. You will find details of my favourite functional testing and how you can get access to this (chapter 19). We've seen how important root cause medicine is, whether optimizing your health or dealing with a diagnosis. This is how you get answers and protect your future.

The MitoImmune Programme is science-led, based on the latest research; the functional testing involves more of an investment and while this isn't available on the NHS in the UK, or may not be covered on insurance in the USA, it does provide invaluable detail into your health. If you decide to go down this route we can support you in doing this (chapter 20). If functional testing is out of your reach for now, the Programme in chapters 14 to 18 is going to provide you with a wealth of health focused steps to bring balance back to your body and protect your future.

So whether you want to implement one change today, or you want to go all in with testing and the full programme, let's get started.

The MitoImmune Nutrition Plan

Why do we start with nutrition?

PEARL: Food is more than fuel, it is a message to every cell in your body. What you eat is telling your body what to do. It's always better to get as much in the way of nutrients as we can from our food. Food truly is medicine.

Nutritional changes aren't the easiest thing to do. They require effort and commitment. I can give you all of the information and knowledge in this book, but if you just read this book then carry on with your life as you are, you won't see any benefits. If you read this book and then implement one, some or all of the steps you will be supporting your health. Don't put pressure on yourself to do everything all in one go, we are all different. I see this in the people I work with. Some want to do it all, right now, take action, go all in and see fast results. Others want to ease in, step by step, slowly changing the way they live. There is no right or wrong. You've purchased this book, you can dip in and out, come back to it when you need and use it as works best for you.

Remember as well that when I work with someone, I fully personalize their nutrition plan to them. We are all unique and should be treated as individuals. You can personalize this programme to work for you, or you can work with a functional medicine practitioner like myself to tailor a plan to you.

I'm a true believer in adding foods as much as possible and only removing foods where necessary, partly because of the wide variety of nutrients our body needs (chapter 11) and partly because food is a large part of our social life (chapter 16). So we are going to start with what you should eat. If you want to tailor this plan to a pescatarian, vegetarian or vegan version, you can do. You will find more meal plans and recipes in the online book resources. If you have an autoimmune condition, such as Hashimoto's for example, you can take the thyroid health steps and the information on autoimmunity in Part 2 and apply that to this plan to adapt it and personalize it to you. Personalization is key and you have the tools to do that here.

What you should eat

1. Whole foods

The MitoImmune Nutrition Plan focuses in on whole foods. Whole foods are packed with fibre and phytonutrients. When foods are refined (i.e. white flour, white bread) much of the protein and fibre is removed leaving the starchy part of the food, meaning they have a bigger impact on your blood sugar and less of the nutrients to benefit the body. The fibre benefits your digestion, gut health and aids in the removal of toxins, cholesterol and other waste products. It also helps support the balance of bacteria in the microbiome. Fibre comes from whole grains, nuts, legumes, vegetables and fruit. We have two types of fibre: insoluble fibre, which is mostly found in the outer skin of vegetables and whole grains, sweeps away debris in your digestive tract and keeps waste moving along; and soluble fibre, which attracts water and provides bulk to your waste, acting as a gel to slow digestion and trap toxins and cholesterol for excretion.

The phytonutrients provide you with a wide mix of nutrients required for all the processes in your body and are highly anti-inflammatory and support oxidative stress in the body. Whole foods also mean you are automatically reducing some of the foods we don't want in the diet such as additives and highly processed foods.

2. Plant foods

It's simple, increase your plant foods. This doesn't have to mean going vegetarian or vegan, but increasing the amount you do eat is going to have huge benefits to all aspects of your health. Plant foods are full of phytonutrients, the compounds that give food their colour, which communicate with the cells in our body and can have a positive impact on how those cells function. Different colours of foods contain different nutrients and we therefore need to eat a wide variety of colours. Fruits and vegetables are rich sources of phytonutrients, along with whole grains, legumes, herbs, spices, nuts, seeds and herbal teas. While darker coloured plant foods are generally higher in phytonutrients, white and tan plant foods also have several beneficial components.

3. Healthy fats

Healthy fats are highly anti-inflammatory. By moving our diet to one that contains more of the healthy fats, we reduce the saturated fats from animal sources and increase omega-3 fatty acids, the anti-inflammatory ones, from fish and plant sources. Oily fish, for example, consumed just one to two times a week, was found in a study to reduce the risk of death from a heart-related problem by up to 36%. Other food sources of healthy fats are free-range eggs, leafy greens, avocado, nuts, seeds and some oils like olive oil. Dark chocolate, 70% or above, is also packed with antioxidants (consume in moderation). Increasing healthy fats is a foundational step to support an anti-inflammatory diet, immune health and brain health. Fats are essential for cell membranes supporting mitochondrial health, energy production and protecting against oxidative stress. Healthy fats also play a supportive role in adrenal health, hormone regulation and thyroid function; they are essential for production of cortisol, support hormone balance by providing the building blocks for hormone production and aiding in the absorption of fat-soluble vitamins, such as vitamin D, which is crucial for hormone regulation and promoting the synthesis and conversion of thyroid hormones.

4. Proteins (adjusted to suit dietary preferences)

Proteins are essential for the repair and maintenance of tissues, including mitochondrial membranes, and provide amino acids necessary for mitochondrial protein synthesis, supporting energy production and brain health via neurotransmitter synthesis and neuroplasticity. Adequate protein intake is crucial for hormone synthesis and regulation, including adrenal and thyroid hormones. Proteins are vital for immune function, as they are involved in the production of antibodies and immune cells, helping to defend the body against infections and diseases. They also contribute to gut health by supporting the integrity of the intestinal barrier and promoting the growth of beneficial gut bacteria. Lastly, protein consumption plays a role in cardiometabolic health by promoting satiety, aiding in weight management and supporting muscle health.

Protein should be eaten at every meal and every snack. Animal proteins are whole proteins, meaning they contain all nine of the essential amino acids – they are essential because we have to consume them, they are not made in the body. Protein from plant sources such as legumes, seeds, nuts and grains usually don't contain all nine amino acids, so it's important to combine foods if getting protein from plant sources to make sure you get all nine amino acids, for example eat legumes or nuts and seeds with rice or grains. Ideally you would eat these combinations at the same meal, but it isn't necessary, so having nuts as a snack and legumes with your dinner makes proteins complete.

5. Legumes and pulses

Legumes and pulses, including beans, lentils, chickpeas and peas, offer many health benefits. They are rich sources of fibre, protein, carbohydrates, vitamins, minerals and phytochemicals, making them valuable components of a balanced diet. They are rich in B vitamins, potassium and magnesium. Consuming legumes and pulses has been associated with improved glycemic control, reduced risk of type 2 diabetes and better weight management, thanks to their low glycemic index and ability to promote satiety. Additionally, the high fibre content of legumes supports gut health. Legumes also contain bioactive compounds, such as polyphenols and flavonoids, which possess antioxidant and anti-inflammatory properties that can help mitigate oxidative stress and reduce inflammation.

6. Nuts and seeds

These are nutrient dense foods that are good sources of healthy fats, protein and are packed full of fibre; they also contain phytochemicals which are similar to plant sterols and help lower cholesterol. They contain key minerals, such as magnesium, selenium and zinc, as well as vitamins like vitamin E. Nuts and seeds provide essential nutrients that support immune function and protect against oxidative stress. These vitamins and minerals play key roles in immune cell function, antibody production and antioxidant defence mechanisms. Additionally, nuts and seeds contain phytochemicals such as flavonoids and polyphenols, which possess anti-inflammatory and immunomodulatory properties, helping to regulate immune responses and reduce the risk of chronic diseases. Nuts and seeds are excellent sources of healthy fats, which have been shown to support mitochondrial function optimizing energy production. Nuts and seeds are also potent sources of nutrients and antioxidants that support brain health, enhancing cognitive function and reducing the risk of age-related cognitive decline. The essential nutrients and healthy fats in nuts and seeds support adrenal and thyroid health, promoting hormone balance and optimal glandular function. They can be eaten as a healthy snack, added to salads, used as a nut butter on fruit, granola or yoghurt, as tahini for a dressing, or added to smoothies.

7. Herbs and spices

Herbs and spices offer a wealth of benefits and have been used as therapeutic foods in some cultures for many years. They contain phytochemicals, antioxidants, vitamins and minerals. Herbs and spices possess anti-inflammatory, antimicrobial, and immune-modulating properties that support overall health and wellbeing. Herbs and spices, such as turmeric, ginger and cinnamon, contain bioactive compounds that enhance mitochondrial function, improve energy metabolism and mitigate oxidative stress, promoting cellular vitality and longevity. They have been shown to support adrenal and thyroid health by modulating stress responses, balancing hormone levels and promoting thyroid function. They also support brain health by improving cognitive function, memory and mood, due to their neuroprotective and cognitive health benefits. They promote digestion, alleviate gastrointestinal discomfort and support gut microbiome balance, benefit blood sugar balance, support lowered cholesterol, reducing cardiovascular risk.

8. Filtered water

Drinking clean filtered water keeps us hydrated, which helps improve metabolism and stress resilience, while removing toxins and benefitting hunger. Reduce or remove alcohol, caffeine and sugary drinks – as these dehydrate, increase cortisol and increase blood sugar – and swap them for filtered water.

In summary

Let's look at how we make these changes in your diet. In what follows you will find action steps to take that will help you to implement the previously mentioned nutrition goals. By doing these action steps you will find that you automatically start to move your diet towards this approach to eating. Take one step at a time, or implement a few or all of them, whichever way feels most comfortable to you. Use the online book resources section as advised to find all the tools you need to support you in making changes.

Make the change – your action steps

Action step 1: phytonutrients

Here is your first challenge; making this one change alone will hugely impact your health. In the online book resources you will find a guide called 'The 6 Colours to Diversity'. In this guide you will see each colour of food, with a list of example foods you could eat for each colour and the benefits that each colour group provides. You will also find 'The Diversity Checklist' – print it off, stick it somewhere you will see it every day and actively use the chart and the tracker.

How to do it?

It's simple, eat two of each colour food daily.

Tick them off the chart.

Action step 2: fruit and vegetable balance

While the UK still recommends five portions of fruit and vegetables a day, some countries have started to increase this. Five is a good start, but the evidence shows we should be eating closer to ten a day. If you have completed action step 1, the phytonutrient chart, you should be doing this, in fact you should be hitting 12. Now we're looking at the balance of where these come from. We are going to break this down into three categories: non-starchy vegetables, starchy vegetables and fruit.

Non-starchy vegetables is where most of your plant points should be coming from – as a minimum you should have six of the ten as non-starchy vegetables. These foods have a wide variety of health benefits and they do not negatively impact your blood sugar. They can lower inflammation, lower blood pressure by relaxing blood vessels, protect the lining of blood vessels and contain vital nutrients for many processes in the body.

The starchy vegetables are rich in phytonutrients and provide carbohydrates, which can benefit energy, but they can spike your blood sugar more than non-starchy vegetables. One to two of these per day is enough for most people, however, athletes or highly active people may require more. The food lists are in the online book resources.

Fruit is full of antioxidants, vitamins, minerals, phytonutrients, fibre and more, but like the starchy vegetables they contain more sugar – it is a natural form of sugar, unlike the processed form in sweet foods. We need to get the nutrients from these foods for our body to function optimally, but we can reduce the impact on your blood sugar by eating fruit with protein or fat, for example, nuts or nut butter. So you may want to add nut butter to your porridge, or have apple slices with nut butter on as a snack, or top your yoghurt and berries with nuts and seeds.

Aim for ten portions a day: six from non-starchy vegetables as a minimum and up to two from starchy vegetables and two from fruit.

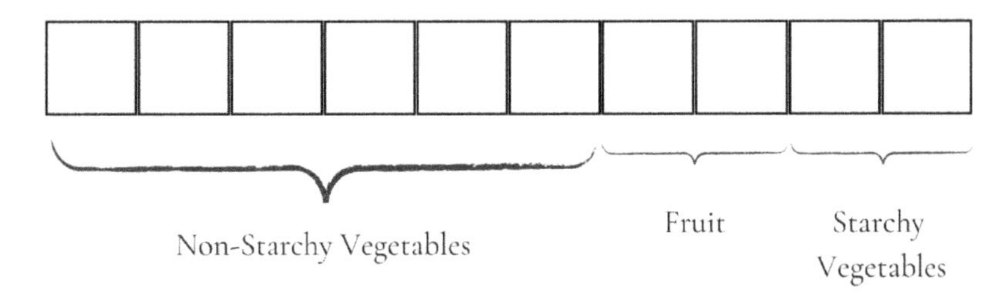

Get the full week printable download in the online book resources, print it off and tick off each box daily.

Action step 3: 50 foods

This is a tool I regularly use with people in clinic. Now that you have built up your phytonutrients you will find it easier to hit this next target. In the online resources you will find your downloadable copy of my 50 foods chart. Your aim? ***Simple hit 50 plant foods in a week.***

Here are the rules:

1. All plant foods can be added to this, but each food only once. So if you eat broccoli on Monday you add it to the chart, if you eat it again on Tuesday that's fine, but you can't add it to the chart again.
2. Different varieties count as different foods, so for example red onion and white onion are two separate foods. Bread and pasta can only go on once if both are made from wheat; if you had wholewheat bread and spelt pasta you can add both as one is wheat and one is spelt.

If you eat 30 in week one, aim for 32 in week two and 34 in week three; grow by two each week until you hit 50.

Action step 4: balanced plate guide

Make sure every plate is well balanced with the different foods. As a guide, at least half of your plate should be vegetables, the more the better. If you can get to three quarters of a plate being non-starchy vegetables even better. A quarter should be taken up by your protein. Then the final quarter split between healthy fats and carbohydrates.

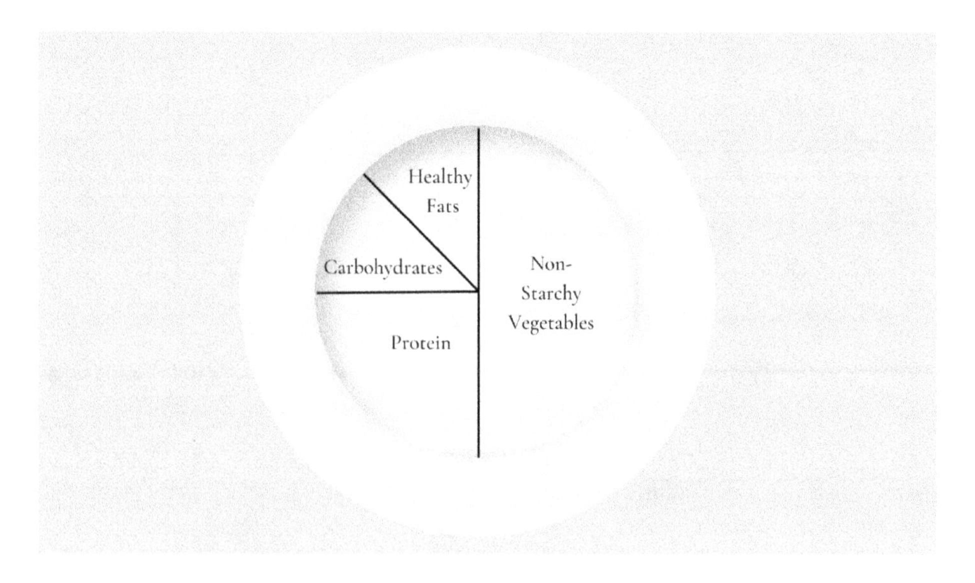

For each plate you build, see if you hit this balance.

Action step 5: balance your proteins

Protein can come from fish, meat or plant-based sources. The type you eat and the quality can mean you have more anti-inflammatory sources or more inflammatory ones; we want to keep the balance pushed to the anti-inflammatory side. Eat protein at every meal and every snack, either animal or plant based. As you've already seen, the more plant foods the better so try to include as much of these as possible.

- ***Aim to eat two–three portions of oily fish per week. Oily fish includes salmon, trout, mackerel, herring, anchovies, sardines.***
- ***Aim to eat two–three portions of lean white meat per week if desired.***
- ***Aim to eat one portion of red meat per week as a maximum, if desired.***
- ***Aim to have a minimum of two days a week where all your protein is from plant-based sources.***

- ***For plant proteins, combine:***
  - ***Beans with grains, nuts or seeds***
  - ***Grains with legumes***
  - ***Nuts or seeds with legumes***
  - ***Vegetables with grains, nuts or seeds***
  - ***Corn with legumes.***

Substitutions:
You can switch this up to work for you and your dietary preferences by doing one of the following:

- You can replace any of these with more plant-based protein days; the more plant-based days the better (if you want to try and go more plant based you will find a meal plan in the online book resources). Your plant-based protein days can be as many as you like.
- If you don't eat red meat you can swap this to a plant-based, or lean white meat instead.

What we don't want you to do is not have adequate protein. Protein helps our bodies to build and repair, without it we can find ourselves lacking in energy, but most people in the western world at least, do not have protein deficiencies.

Note:
The quality of these foods is vital, and we talk about this more in chapter 17. While working on this step make sure you gain an understanding on how to choose the right quality of these foods. I would rather you ate one portion of really good quality red meat a month than poor quality red meat once a week.

Action step 6: anti-inflammatory foods

The MitoImmune Nutrition Plan is anti-inflammatory in nature, but here we will focus on some specific nutrients, strongly correlated with reduced inflammation in the body.

Foods of particular use due to their strong correlation with being anti-inflammatory are:

- Pterostillbene found in blueberries
- Curcumin, the active ingredient found in turmeric
- Quercetin, found in apples, any dark coloured berries, broccoli, cauliflower, cabbage, sprouts, olive oil, capers and onions
- Carotenoids and flavonoids, found in dark leafy green and cruciferous vegetables such as kale, spinach, bok choy, watercress, broccoli and chard
- Beta-Carotene, found in carrots, mangoes, orange and yellow peppers
- Bromelain, an anti-inflammatory enzyme found in pineapple
- Carnasol, found in rosemary.

Aim to include at least three of these a day.

Anti-inflammatory teas:

- Ginger, lemon and honey – an inch cut of ginger, with a slice of lemon and half a teaspoon of honey in a mug with hot water.
- Turmeric Milk – 2tsp turmeric, 2 tsp cinnamon, 2 tsp maple syrup (optional), a sprinkling of black pepper, mixed with 600ml hot almond milk, heated in a pan. (I like to use my milk whisker to add some frothy milk on top.) Serves 2.
- Chamomile tea.
- Green tea.
- Peppermint tea.

Aim to include two–three cups of anti-inflammatory tea a day.

Fermented foods:[3]
The fermentation process makes enzymes that break down some protein and lactose, the sugar in dairy, making them easier to digest; they also have nutritional benefits in vitamins like B, K, and they produce by-products as the fermentation process happens, which can fight inflammation.

Foods to include are kefir, sauerkraut, kimchi and yoghurt.

Action step 7: herbs and spices

We've looked at the health benefits of herbs and spices so your next step is to increase their intake in your diet.

Aim to include two herbs or spices at each meal.

People often find breakfast the hardest meal to do this. If you have eggs you can look to include herbs like dill and parsley in an omelette or scrambled eggs. If you have porridge think about adding cinnamon or nutmeg. Alternatively you could try a turmeric latte with your breakfast.

Action step 8: flaxseed

Aim to add 2 tablespoons of flaxseed to your breakfast, yoghurt or smoothie daily.

If new to increasing fibre, you can build this up slowly and make sure you stay hydrated when increasing fibre.

[3] Avoid if you have problems with histamine.

Action step 9: hydrate

Aim to drink 2.5–3 litres of water a day.

Herbal teas (if decaffeinated) can count towards this amount, plus broths, but coffee and black tea don't. If you don't like drinking water and want to avoid sugary drinks, try adding a slice of lemon or lime, fresh fruit or a splash of fresh natural juice.

Foods to remove

Now that you have taken steps to increase foods in the diet, let's spend some time on foods you should reduce, or eliminate. As I've said I'm not an advocate of any one food group being 'bad' and needing to be removed from everyone's diet, what I am an advocate of is removing foods that are not whole foods, those which are highly processed, or refined and which promote inflammation.

Action step 10: reduce or remove gluten

Focusing on The MitoImmune Nutrition Plan means you will be focusing naturally on whole, unprocessed foods such as fruits, vegetables, lean proteins, nuts and seeds. Now it's time to look at swapping out gluten-containing grains like wheat, barley and rye for alternative gluten-free grains such as quinoa, rice, millet and buckwheat. Experiment with gluten-free flours like almond flour, coconut flour and tapioca flour for baking and cooking. Additionally, explore gluten-free alternatives for common staples such as bread, pasta and snacks, which are widely available in most grocery stores. Try lentil or green pea pasta, for example. What we don't want you to do is go gluten free and hit the free-from isle and buy lots of processed gluten-free foods. Instead, think about focusing on the naturally gluten-free foods.

Gluten-containing grains:

- Wheat
- Barley
- Rye.

Alternative gluten-free grains:

- Quinoa
- Rice
- Millet
- Buckwheat
- Oats.

If you are struggling to reduce or remove gluten, head to the online book resources where you can get access to my gluten-free meal plan and recipes, plus guides to support you on this journey.

Action step 11: reduce sugar

We all have some sugar sometimes, whether it's a slice of cake or a bar of chocolate; of course there are options to reduce sugar load here. A patient recently asked me what cake she could make that would be a healthier option. She was tasked with taking the dessert to a friend's get together. I told her that she could try a cake made with ground almonds instead of white flour and using dark chocolate, more than 70%. Less refined white flour, less sugar and the benefit of nuts plus antioxidants from the dark chocolate, and she still got to show up with a chocolate cake, dig in and enjoy with everyone else.

Now, of course, even with the swaps, we don't want you eating cake or chocolate, for example, every day. What this does show you though is that you can take steps to lighten the load. It's not about perfection. If a friend serves you dessert, live, enjoy and don't beat yourself up, but reduce your sugar intake at home.

The impact of sugar:

- Weight gain and obesity
- Blood sugar balance
- Inflammation linked to heart disease, strokes, autoimmune disorders and more
- Negatively impacts cognitive function, mood regulation and brain health
- Impaired or suppressed immune function
- Dysregulation of the hunger hormones
- Triggers the release of stress hormones, such as cortisol and adrenaline (epinephrine); continued high sugar intake can contribute to the development of adrenal fatigue
- Disruption of TSH, T4 and T3 balance and impairing conversion of T4 to T3.

Aim to reduce sugar as much as possible, and where required use some natural sweeteners in moderation such as molasses, honey, maple syrup and coconut sugar. Moderation being key. If you use sugar in your tea you can start by switching this for honey, let the tea cool to drinkable temperature and then add the honey.

If you are struggling to go sugar free, or are having cravings, head to the online book resources where you can get access to my sugar-free meal plan and recipes, plus guides on snacking and cravings to support you.

Action step 12: reduce processed foods

Reducing processed foods in the diet is paramount for promoting overall health and wellbeing, as these products often contain artificial colourings and flavourings, additives, preservatives and harmful fats such as trans fats, hydrogenated fats and partially hydrogenated fats. Artificial colourings and flavourings, commonly found in processed foods, have been linked to hyperactivity in children, allergic reactions and adverse effects on behaviour and cognition. Additionally, additives and preservatives, such as sulfites, nitrates and benzoates, can trigger allergic reactions and may have carcinogenic properties. Moreover, trans fats, hydrogenated fats and partially hydrogenated fats found in processed foods can increase levels of LDL cholesterol, lower levels of HDL cholesterol and promote inflammation, significantly raising the risk of cardiovascular disease. These harmful fats also impair mitochondrial function, leading to reduced energy production and increased oxidative stress, contributing to fatigue and cellular damage. Furthermore, the consumption of processed foods can negatively impact immune function by promoting inflammation, disrupting gut microbiota balance and compromising the integrity of the gut barrier, making individuals more susceptible to infections and autoimmune diseases. The excessive consumption of processed foods has been associated with poor brain health, cognitive decline and an increased risk of neurodegenerative diseases, such as Alzheimer's disease, due to the detrimental effects of artificial additives and unhealthy fats on neuronal function and synaptic plasticity. Additionally, processed foods are typically low in fibre and essential nutrients, further compromising gut health and contributing to digestive issues such as constipation, bloating and irritable bowel syndrome. Overall, reducing processed foods in the diet is crucial to support optimal health.

Action step 13: reduce alcohol

Alcohol is a calorie-dense substance that provides little to no nutritional value and can contribute to weight gain and obesity when consumed in excess. Excessive alcohol consumption is associated with an increased risk of liver damage, including fatty liver disease, hepatitis and cirrhosis, as alcohol is metabolized primarily in the liver. Alcohol can weaken the immune system, making individuals more susceptible to infections and illnesses and can have both short-term and long-term impacts on the brain. In the short term, alcohol consumption can impair cognitive function, coordination and decision-making abilities, and longer-term alcohol consumption can damage brain cells, disrupt neurotransmitter systems and impair synaptic plasticity, leading to cognitive deficits and difficulties with learning and memory. It can exacerbate mental health disorders such as depression and anxiety, leading to worsened symptoms and decreased overall wellbeing.

Moderate alcohol consumption, specifically of red wine, has been associated with certain health benefits, such as a reduced risk of heart disease and stroke when consumed in moderation. However, moderation is key, as excessive alcohol intake negates any potential benefits and increases the risk of adverse health outcomes.

Red wine

The primary antioxidants found in red wine are polyphenols, particularly a group called flavonoids, which include resveratrol, quercetin and catechins. These antioxidants are mainly found in the skins and seeds of grapes, and during the winemaking process, the fermentation of red wine includes more contact with grape skins, leading to a higher concentration of polyphenols compared to white wine.

Resveratrol, in particular, has garnered attention for its potential health benefits, including its antioxidant and anti-inflammatory properties. Some research suggests that moderate consumption of red wine, due to its higher polyphenol content, may offer certain health benefits, such as reducing the risk of cardiovascular disease and improving blood lipid profiles.

To optimize alcohol consumption, enjoy a clear spirit drink with a lower sugar mixer or alone; or a glass of wine, preferably red for the polyphenols – drink in moderation and consider having alcohol with an appetiser like nuts, which can offer certain benefits. When consumed with alcohol, the healthy fats in nuts can help to slow down the absorption of alcohol in the bloodstream, potentially reducing the rate at which blood alcohol levels rise. Organic wine is also a good choice – as with food, organic wine is made from grapes grown without the use of synthetic pesticides, herbicides or fertilisers. Read more about the benefits of going organic in chapter 17.

Action step 14: reduce caffeine

I'm not someone who will demonize caffeine, in fact it has benefits. I love a coffee in the morning but I don't drink it all day. I make sure I also drink enough water.

I rarely have it after lunchtime, if ever, and I don't add sugar. Caffeine acts on the central nervous system, blocking the action of adenosine, a neurotransmitter that promotes relaxation and sleepiness; having caffeine too late can impair sleep and the half-life of coffee is quite long – this is the time it takes for half of the caffeine to be eliminated from the body, therefore not having coffee after lunch is a good boundary to have. Caffeine stimulates the release of adrenaline, a hormone that increases heart rate and blood pressure. Excessive caffeine intake, or sensitivity to caffeine, can lead to palpitations, rapid heartbeat and elevated blood pressure.

Some research suggests that moderate caffeine intake may actually be associated with a reduced risk of neurodegenerative diseases such as Alzheimer's and Parkinson's disease. Caffeine's neuroprotective effects may be attributed to its antioxidant properties and ability to modulate neurotransmitter systems involved in brain health.

Eating behaviours

Intermittent fasting

When we talk about intermittent fasting there are many types that are discussed. The type of intermittent fasting I often work with is time-restricted eating – this type of intermittent fasting is an eating pattern that involves limiting the window of time during which food is consumed each day, while fasting for the remainder of the day and overnight. This way of eating, unlike a diet, doesn't focus on restricting food groups or calorie intake, instead it primarily dictates when to eat, rather than what or how much to eat. It is not only simple to implement, it also has many health benefits.

Time-restricted eating[4] revolves around establishing a daily eating window. The scientific studies have typically ranged from a 4- up to 12-hour window, during which all meals and snacks are consumed. The remaining hours of the day, known as the fasting window, are dedicated to abstaining from food, allowing the body to rest and reset.

The concept of time-restricted eating is rooted in the idea that the human body operates on a circadian rhythm, a 24-hour cycle that regulates various physiological processes, including metabolism, hormone production and sleep–wake cycles. By aligning eating patterns with the body's natural circadian rhythms, it has been seen to optimize metabolic function, promote weight loss and improve overall health.

---

[4] Despite its potential benefits, time-restricted eating may not be suitable for everyone, particularly those with certain medical conditions, such as diabetes, eating disorders or hormonal imbalances. Always speak to a practitioner for personalized advice on whether it may work for you.

During the fasting period, the body undergoes several metabolic changes that contribute to its health benefits. In the absence of incoming nutrients, insulin levels decrease, signalling the body to switch from using glucose as its primary fuel source to burning stored fat for energy, a process known as ketosis. This shift in metabolism can lead to increased fat burning and weight loss, particularly when combined with regular physical activity. Fasting also triggers a cellular repair process called autophagy, in which the body breaks down and recycles damaged, or dysfunctional, cellular components. Autophagy helps remove toxins, repair DNA damage and promote cellular renewal, which has been shown to have anti-ageing and disease-preventive effects.

Breakfast like a king

There is also a theory, again based in our circadian cycle, that eating breakfast like a king, lunch like a prince and dinner like a pauper may be beneficial for health. The idea being that you should have the larger amount of calories when your metabolism is typically more active and you have the day ahead to burn. It involves having your largest meal at breakfast, a mid-sized meal for lunch and a smaller dinner. Eating a substantial breakfast provides the body with the necessary fuel and nutrients to kickstart metabolism, stabilize blood sugar levels and promote satiety, reducing the likelihood of overeating later in the day. A balanced lunch provides sustained energy and nutrients to support cognitive function, productivity and physical activity throughout the afternoon. By consuming a lighter dinner, the body can focus on digestion and repair during the overnight fast, promoting restful sleep and facilitating cellular repair and renewal.

Eating on the run

In today's modern, fast-paced lifestyle, eating on the run has become a common practice for many of us. Driven by busy schedules, work demands and constant connectivity, we no longer savour meal times and take a proper break when eating. However, this hurried approach to eating can have significant consequences for digestion and overall health. Digestion is a complex process that begins even before food enters the mouth. The mere smell, or sight, of food triggers the release of saliva. When we take time to cook a meal the process of digestion starts as we prepare our meal. This allows the production of saliva, which contains enzymes that break down our food. However, when meals are rushed and consumed hastily, this crucial initial phase of digestion may be compromised. We then don't chew our food enough as we are eating in a rush to get back to our busy days. Inadequate chewing of food can hinder the mechanical breakdown of food particles into smaller, more digestible pieces, placing greater strain on the digestive system downstream. This can lead to digestive discomfort, such as bloating, gas, and indigestion, as well as impaired nutrient absorption and utilization. To optimize digestion and support overall health, it is essential to slow down, take time over meals and chew food thoroughly. Mindful eating practices, such as paying attention to the sensory experience of eating, savouring each bite and chewing food slowly and deliberately, can enhance digestion, promote satiety and foster a greater connection with food. Additionally, eating in a relaxed environment, free from distractions, can help activate the parasympathetic nervous system, which is responsible for rest and digest functions. By prioritizing mindful eating habits, such as 3-4-5 breathing before eating (see chapter 16) and slowing down the pace of meals, you can support optimal digestion, nutrient absorption and overall wellbeing in today's fast-paced world.

15

Specific nutrients

The MitoImmune Nutrition Plan is designed to support all the pillars of your health, but as we have seen in Part 2, some nutrients are particularly targeted to specific pillars – you will find a list at the end of each chapter in Part 2 to refer to. So if you want to dive deeper into nutritional changes and you feel you have implemented the steps in The MitoImmune Nutrition Plan, this is your next step to dig deeper. Remember, if you are not sure which pillars to focus on, take The MitoImmune Health Assessment in the online book resources first to assess each pillar for you personally, then start with the lowest scoring.

Alpha-lipoic acid (ALA)

ALA is a powerful antioxidant and helps with free radicals and oxidative stress, benefiting all of the pillars of health, especially mitochondrial, adrenal, immune, brain, cardiometabolic and thyroid health. ALA is also a cofactor for several enzymes involved in mitochondrial energy production promoting cellular energy production. The antioxidant properties help reduce inflammation, which can support a healthy immune response. By neutralizing harmful free radicals, ALA may help strengthen the immune system and enhance its ability to combat infections and pathogens. ALA has been shown to cross the blood-brain barrier and exert neuroprotective effects in the brain. ALA may support cognitive function, memory and overall brain health. Some research suggests that ALA may be beneficial for conditions such as Alzheimer's disease and age-related cognitive decline. ALA's antioxidant and anti-inflammatory properties may help support gut health by reducing inflammation and damage to intestinal cells. By protecting against gut inflammation and dysfunction, ALA may help support a healthy gut microbiome and improve digestive function.

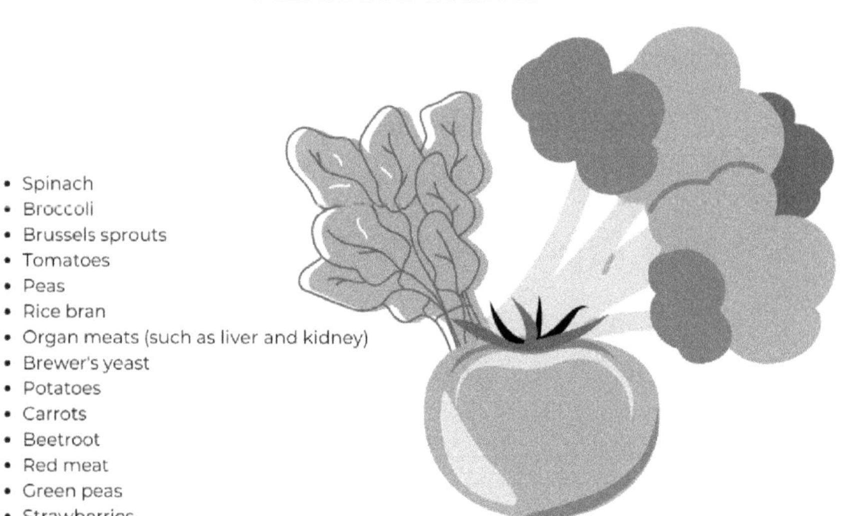

Brain-derived neurotrophic factor (BDNF)

BDNF is a protein which helps to build and protect nerve cells and neurons in the brain. It has been shown to be associated with higher cognitive and brain performance and important for thinking and learning. Lowered levels of BDNF have been associated with multiple sclerosis, Parkinson's and Alzheimer's disease. Lowering blood sugar, higher omega-3 intake, especially DHA, curcumin, resveratrol and polyphenols (which we will come on to shortly) and intermittent fasting have been shown to increase levels. Lifestyle factors (chapter 16) such as increased sunshine, meditation, lowered stress, improved sleep quality, movement and mental stimulation have also been shown to play an important role in increasing BDNF levels.

B vitamins

The B vitamins, including B1 (thiamine), B2 (riboflavin), B3 (niacin), B5 (pantothenic acid), B6 (pyridoxine), B7 (biotin), B9 (folate) and B12 (cobalamin), play essential roles in numerous physiological processes and are crucial for overall health and wellbeing. They play important roles in energy production and increase the activity of the mitochondria. They are also protective against oxidative stress (B2) and have been shown to improve cognitive function and brain performance. B1 aids in converting carbohydrates into energy; B2, B3 and B5 participate in the production of ATP (B5 is also involved in the synthesis of coenzyme A (CoA), which is necessary for the metabolism of fats, carbohydrates and proteins); B7 is also beneficial here. B12 is required for conversion of fats and proteins into energy. Certain B vitamins also provide targeted support for the adrenals, especially B5 and B6.

B1 is also important for nerve function and transmission, supporting proper brain and nervous system health. B12 supports cognitive function and prevention of nerve damage. B3 plays a role in DNA repair and synthesis, supporting cellular health and integrity and B9 is essential for DNA synthesis and cell division, supporting growth and development, B12 is also important for DNA synthesis. B6 is essential for the synthesis of neurotransmitters such as serotonin, dopamine and GABA, which are involved in mood regulation and cognitive function.

B vitamins, particularly B2, B3, B6 and B12, are involved in various metabolic processes that support thyroid function. They help convert inactive thyroid hormone (T4) into its active form (T3) and support energy metabolism.

B1 and B12 help maintain healthy heart function and cardiovascular health. B2 is essential for the metabolism of fats, carbohydrates and proteins. B3 helps maintain healthy cholesterol levels by raising HDL (good) cholesterol and lowering LDL (bad) cholesterol and triglycerides. B7 is supportive in glucose metabolism, benefiting blood sugar levels.

B6 and B9 are important for immune function and may help support lowered inflammation, while B5 is important for adrenal function, stress and fatigue. B2, B5 and B7 are important for maintaining healthy hair, eyes, nails and skin. B6 and B12 are also beneficial in hormone balance – they play essential roles in hormone metabolism and regulation. They help support the synthesis and metabolism of hormones such as oestrogen, progesterone and testosterone.

B Vitamin Food Sources

- B1 (Thiamine):
  - Whole grains (such as brown rice, whole wheat, oats)
  - Legumes (such as lentils, beans)
  - Nuts and seeds (such as sunflower seeds, peanuts)
  - Pork
  - Yeast extract (such as nutritional yeast)
- B2 (Riboflavin):
  - Dairy products (such as milk, yogurt, cheese)
  - Eggs
  - Lean meats (such as chicken, turkey)
  - Leafy green vegetables (such as spinach, kale)
  - Mushrooms
  - Almonds
- B3 (Niacin):
  - Meat (such as beef, chicken, fish)
  - Whole grains (such as brown rice, barley)
  - Legumes (such as peanuts, lentils)
  - Seeds (such as sunflower seeds)
  - Leafy green vegetables (such as spinach)
  - Mushrooms
  - Avocado
- B5 (Pantothenic Acid):
  - Meat (such as beef, pork, chicken)
  - Fish (such as salmon, tuna)
  - Whole grains (such as whole wheat, oats)
  - Legumes (such as lentils, chickpeas)
  - Eggs
  - Avocado
- B6 (Pyridoxine):
  - Poultry (such as chicken, turkey)
  - Fish (such as salmon, tuna)
  - Potatoes
  - Bananas
  - Legumes (such as chickpeas, lentils)
  - Whole grains (such as brown rice, barley)
- B7 (Biotin):
  - Egg yolks
  - Liver
  - Nuts and seeds (such as almonds, peanuts)
  - Legumes (such as lentils)
  - Whole grains (such as oats)
  - Spinach
- B9 (Folate):
  - Leafy green vegetables (such as spinach, kale)
  - Legumes (such as lentils, chickpeas)
  - Citrus fruits (such as oranges, grapefruit)
  - Avocado
  - Asparagus
  - Beets
- B12 (Cobalamin):
  - Meat (such as beef, pork, lamb)
  - Fish (such as salmon, trout, tuna)
  - Shellfish (such as clams, mussels)
  - Dairy products (such as milk, cheese, yogurt)
  - Eggs
  - Fortified foods (such as cereals, plant-based milk)

Carnitine

Carnitine, a naturally occurring compound synthesized from the amino acids lysine and methionine, plays a vital role in energy metabolism and cellular function. It helps transport fats into the mitochondria so that they can be used in the production of energy. Improving carnitine levels may benefit fatigue and brain function. Carnitine has also been linked to improved exercise performance, muscle recovery and endurance. There is also some research into carnitine for brain health and cognitive performance but we need more scientific evidence of this impact.

Carnitine Food Sources

Carotenoids

Found predominantly in yellow, orange and red fruit and vegetables, carotenoids are potent antioxidants helping to neutralize free radicals, reducing oxidative stress and inflammation. They support immune function, brain health and cognitive function.

Carotenoid Food Sources

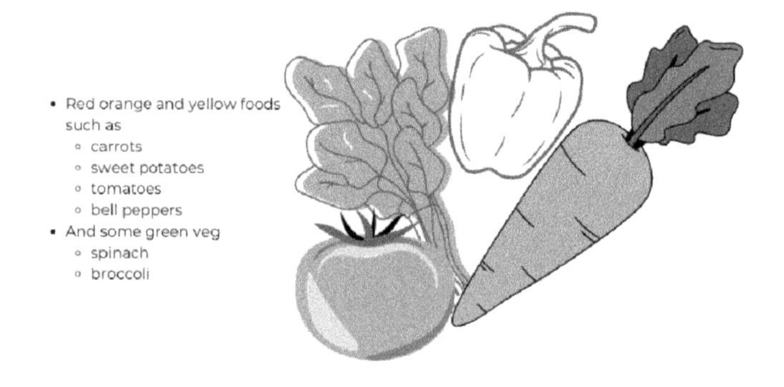

Choline

Choline is a precursor to acetylcholine, a neurotransmitter involved in muscle control, memory and cognitive function. Choline is vital for brain health and cognitive development. Choline is a key component of cell membranes and is involved in the production of phospholipids, which are essential for cellular structure and integrity. Choline also plays a role in lipid metabolism, helping to transport fats from the liver and regulate cholesterol levels. Moreover, choline has been linked to improved performance, muscle function and recovery, as well as reduced inflammation and oxidative stress.

Choline Food Sources

Coenzyme Q10 (CoQ10)

The anti-inflammatory and antioxidant CoQ10 supports mitochondrial function by protecting against oxidative stress and cellular energy production. It is a naturally occurring compound found in every cell of the body. It is particularly abundant in organs with a high energy demand, such as the brain and heart, and is vital for optimizing energy levels.

CoQ10 Food Sources

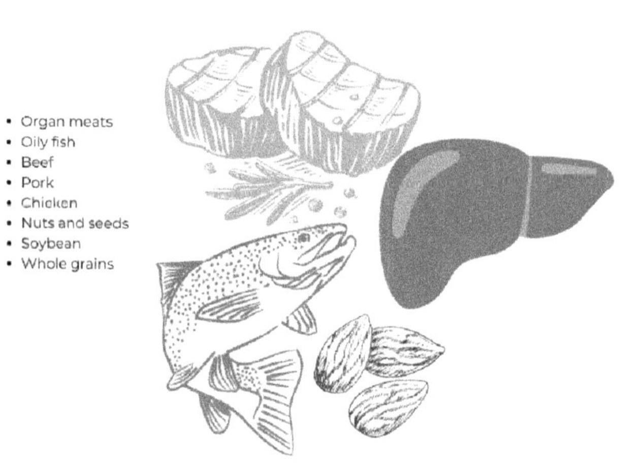

Flavonoids

Flavonoids are a diverse group of plant compounds found in fruits, vegetables, tea, cocoa and red wine, known for their potent antioxidant and anti-inflammatory properties. They offer a wide range of health benefits. They help reduce the risk of heart disease by improving blood flow, lowering blood pressure and reducing inflammation and oxidative stress. Flavonoids also support cognitive function and brain health by enhancing memory, learning and cognitive flexibility, and may reduce the risk of age-related cognitive decline and neurodegenerative diseases such as Alzheimer's. They support immune function by modulating immune responses and reducing inflammation, helping the body defend against infections and diseases.

Flavonoid Food Sources

- Berries:
  - Blueberries, strawberries, raspberries, blackberries and cranberries are particularly rich sources of flavonoids, including anthocyanins.
- Citrus fruits:
  - Oranges, lemons, limes and grapefruits contain flavonoids such as hesperidin and naringin.
- Apples:
  - Apples contain flavonoids such as quercetin and catechins, particularly in the skin.
- Grapes:
  - Grapes and grape products (such as red wine and grape juice) contain flavonoids such as resveratrol and quercetin.
- Tea:
  - Green tea and black tea are rich sources of flavonoids, particularly catechins and theaflavins.
- Dark chocolate:
  - Dark chocolate contains flavonoids called flavanols, particularly epicatechin.
- Red wine:
  - Red wine contains flavonoids such as resveratrol, particularly in the skin of red grapes.
- Leafy greens:
  - Dark leafy greens such as spinach, kale and Swiss chard contain flavonoids such as quercetin and kaempferol.
- Onions:
  - Onions, particularly red onions, contain flavonoids such as quercetin.
- Soy products:
  - Soybeans and soy products contain flavonoids such as isoflavones, particularly genistein and daidzein.

Glutamine

Glutamine is the most abundant amino acid in the body and is particularly important for supporting gut health, immune function and muscle repair. It is a prime source of fuel for the intestinal cells and supports the integrity and function of the gastrointestinal tract, helping to maintain the gut barrier and prevent, or heal, intestinal permeability. Glutamine also serves as a primary energy source for immune cells, supporting their proliferation and function, which is essential for mounting an effective immune response against pathogens and infections.

Glutamine Food Sources

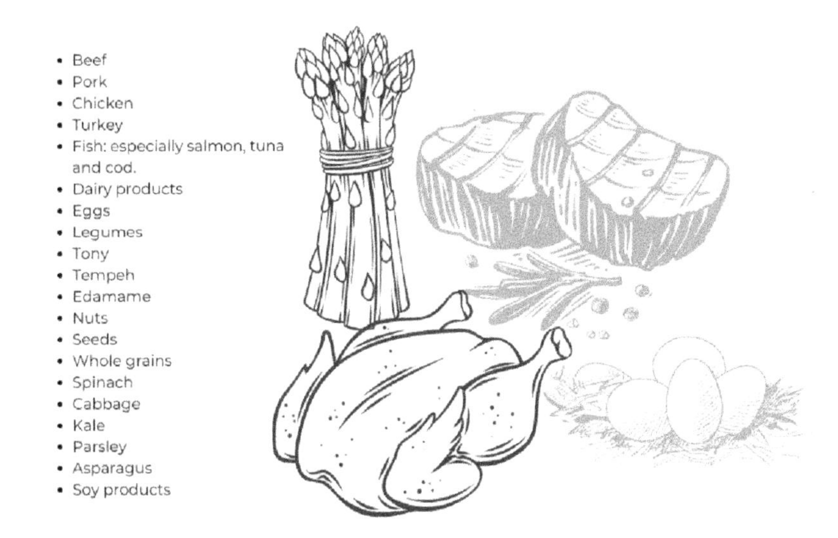

Glutathione (or its precursor N-Acetyl-Cysteine (NAC))

Glutathione, often referred to as the body's 'master antioxidant', plays a crucial role in protecting cells from oxidative stress and damage caused by free radicals. As a powerful antioxidant, glutathione helps neutralize harmful reactive oxygen species (ROS) and toxins, thereby reducing inflammation and preventing cellular damage. This is essential for optimal health and longevity. Glutathione supports detoxification processes in the liver by binding to and eliminating harmful substances such as heavy metals, pollutants and carcinogens. Additionally, glutathione plays a vital role in immune function by supporting the activity of white blood cells and enhancing the body's ability to fight infections and diseases. Moreover, glutathione has been linked to improved energy levels, enhanced cognitive function and reduced risk of chronic diseases such as cancer, heart disease and neurodegenerative disorders.

Glutathione Food Sources

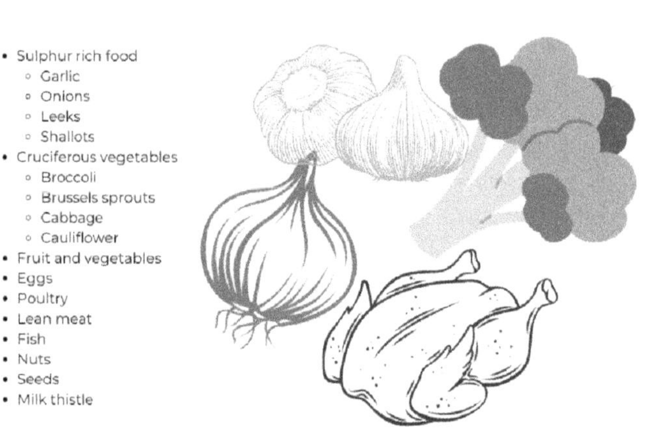

Iron

Iron is a key component of haemoglobin, a protein found in red blood cells that binds to oxygen in the lungs and carries it to tissues and organs throughout the body. Adequate iron levels are necessary for maintaining optimal oxygen delivery to cells, which is essential for energy production and overall vitality. Iron is required for the production of ATP. Iron-containing enzymes, such as those involved in the electron transport chain in mitochondria, help facilitate the conversion of nutrients into usable energy. Iron plays a role in immune function and helps support the

Iron Food Sources

body's defence against infections and diseases. It is required for the proliferation and function of immune cells, including lymphocytes and macrophages, which help identify and eliminate pathogens such as bacteria, viruses and fungi. Iron is necessary for the production of thyroid peroxidase, an enzyme involved in the synthesis of thyroid hormones – low iron can contribute to thyroid dysfunction. Iron is important for cognitive function and brain health. Iron is involved in the production of neurotransmitters such as dopamine, serotonin and norepinephrine, which play key roles in mood regulation, cognitive function and behaviour and memory.

Magnesium

Magnesium is a cofactor in numerous enzymatic reactions involved in ATP production and mitochondrial function. It helps regulate mitochondrial calcium levels and supports mitochondrial stability. Magnesium is also involved in numerous biochemical processes in the brain and supports cognitive function, mood regulation and sleep quality. Magnesium can become depleted by stress and it supports neurotransmitter production that benefits feeling calm. It plays a role in hormone regulation and supports the function of the endocrine system as well as playing a crucial role in maintaining cardiovascular health. It helps regulate heart rhythm, blood pressure and blood vessel function. Magnesium may help reduce the risk of cardiovascular diseases such as heart disease, stroke and hypertension, and it helps regulate blood sugar levels by facilitating the uptake of glucose into cells and promoting insulin secretion. Adequate magnesium levels may help improve insulin sensitivity and reduce the risk of type 2 diabetes. Magnesium is involved in the regulation of the sleep–wake cycle and may help improve sleep quality. It helps relax muscles and promote feelings of relaxation and calmness, which are conducive to falling asleep and staying asleep.

Magnesium Food Sources

Omega 3

Omega-3 fatty acids, particularly eicosapentaenoic acid (EPA) and docosahexaenoic acid (DHA), support mitochondrial function by enhancing mitochondrial membrane fluidity and reducing inflammation. DHA, in particular, is highly concentrated in the brain and plays a crucial role in brain development and function. Omega-3 fatty acids support cognitive function, memory and learning, and may help reduce the risk of age-related cognitive decline and neurodegenerative diseases such as Alzheimer's disease and dementia. They also support mood regulation and may help reduce symptoms of depression and anxiety. They support the function of immune cells and enhance the body's ability to combat infections. The reduction in inflammation is also beneficial for the adrenals and hormone production. They help reduce triglyceride levels, lower blood pressure and decrease inflammation in the body, all of which contribute to a reduced risk of heart disease and stroke. Omega-3 fatty acids may also help prevent the formation of blood clots and plaque build up in the arteries.

Omega-3 Food Sources

- Oily fish: salmon, mackerel, herring, anchovies, trout and sardines
- Flaxseeds
- Chia seeds
- Walnuts

Polyphenols

Polyphenols are plant compounds with antioxidant and anti-inflammatory properties that may support brain health and cognitive function among other things. They help protect brain cells from oxidative damage, reduce inflammation and support neuroplasticity. Let's look at some of the more potent specific polyphenols.

Berberine

Berberine has been studied for its potential to improve metabolic health by supporting blood sugar regulation and insulin sensitivity. Berberine may help support heart health by reducing cholesterol levels, triglycerides and blood pressure. It may also help improve lipid metabolism and reduce inflammation in the arteries. Berberine has antimicrobial properties and may help support a healthy balance of gut microbiota. It may also help reduce intestinal inflammation and support digestive health. Berberine is usually given as a supplement.

Curcumin

Curcumin is well-known for its potent anti-inflammatory properties. It helps reduce inflammation in the body by inhibiting inflammatory pathways and reducing the production of inflammatory molecules. Curcumin is a powerful antioxidant that helps protect cells from oxidative damage caused by free radicals. Curcumin may help support brain health and cognitive function by crossing the blood-brain barrier and exerting neuroprotective effects. It may help reduce inflammation in the brain and promote the growth of new neurons.

Curcumin Food Sources

- Turmeric
- Curry powder
- Mustard
- Yellow mustard seeds

Resveratrol

Resveratrol has been studied for its potential cardiovascular benefits. It may help improve heart health by reducing inflammation, lowering blood pressure and improving blood vessel function. It may also help protect against atherosclerosis and reduce the risk of heart disease. Resveratrol has been investigated for its potential anti-ageing effects. It may help activate longevity genes and promote cellular repair mechanisms, leading to improved cellular function and longevity. Resveratrol may support brain health and cognitive function by reducing inflammation, promoting neuroplasticity and protecting against neurodegenerative diseases such as Alzheimer's and Parkinson's.

Resveratrol Food Sources

- Red grapes
- Red wine (in moderation)
- Peanuts
- Pistachios
- Blueberries
- Cranberries
- Dark chocolate
- Red onions

Quercetin

Quercetin has immunomodulatory properties and may help support immune function. It has been studied for its potential to reduce inflammation, enhance antioxidant defenses and regulate immune cell activity. Quercetin may help support heart health by reducing inflammation, improving blood vessel function and lowering blood pressure. It may also help protect against oxidative damage and reduce the risk of cardiovascular diseases.

Quercetin Food Sources

- Apples (with skin)
- Onions (red onions are particularly high)
- Garlic
- Dark berries (such as cranberries, blueberries and raspberries)
- Citrus fruits
- Leafy greens
- Tomatoes
- Red grapes
- Broccoli
- Green tea

Phospholipids

Phospholipids, particularly phosphatidylserine (PS) and phosphatidylcholine (PC), are important components of cell membranes in the brain. Phospholipids are the primary building blocks of cell membranes, forming a lipid bilayer that surrounds and protects cells. They help maintain the structural integrity and fluidity of cell membranes, allowing for the passage of nutrients and waste products in and out of cells. Phospholipids help form myelin sheaths, the protective covering of nerve fibres that facilitate efficient nerve impulse transmission. They also support neurotransmitter synthesis and release, which are essential for memory, learning and overall cognitive function.

Phospholipid Food Sources

Pre and probiotics

Probiotics are beneficial bacteria that help maintain a healthy balance of gut microbiota. They support digestion and nutrient absorption. Probiotics support immune function by maintaining a healthy balance of gut microbiota and enhancing the body's defence against pathogens. Prebiotics are non-digestible fibres that promote the growth and activity of beneficial gut bacteria. They help nourish the gut microbiome and support digestive health. If you think of probiotics as the soil for the plants, the prebiotics are the fertiliser you add in.

Pre- & Probiotic Food Sources

- Probiotic food sources:
  - Yogurt
  - Kefir:
  - Fermented Vegetables
  - Miso
  - Tempeh
  - Kombucha
- Prebiotic food sources:
  - Chicory Root
  - Garlic
  - Onions
  - Jerusalem Artichoke
  - Bananas
  - Asparagus
  - Apples

YOGURT

Kombucha

Pyrroloquinoline Quinone (PQQ)

PQQ is a potent antioxidant helping to neutralize harmful free radicals and reduce oxidative stress in the body. By scavenging free radicals, PQQ helps protect cells and tissues from damage caused by oxidative stress, which is implicated in ageing, chronic diseases and inflammation. PQQ plays a crucial role in mitochondrial function and biogenesis – it helps optimize mitochondrial function, enhance energy production and improve cellular metabolism. PQQ has been shown to exert neuroprotective effects by promoting the growth and survival of nerve cells and protecting against oxidative damage and neurotoxicity. PQQ may help support brain health, cognitive function and memory by reducing the risk of age-related cognitive decline and neurodegenerative diseases such as Alzheimer's and Parkinson's.

PQQ may also help support cardiovascular health by reducing inflammation, improving blood vessel function and protecting against oxidative stress and lipid peroxidation. PQQ has been shown to lower blood pressure, improve cholesterol levels and reduce the risk of atherosclerosis and heart disease. PQQ has anti-inflammatory properties and helps modulate immune responses by inhibiting inflammatory pathways and reducing the production of pro-inflammatory cytokines.

PQQ Food Sources

- Parsley (Parsley is one of the richest sources of PQQ among common foods. Adding fresh parsley to salads, soups and dishes can boost your PQQ intake)
- Bell peppers
- Green Tea
- Tofu
- Natto (Natto is a traditional Japanese dish made from fermented soybeans and is known for its distinct flavour and texture.)
- Kiwi
- Spinach

Selenium

Selenium is a trace mineral that supports immune function by enhancing the activity of immune cells and protecting against oxidative stress. Selenium is required for the production and function of immune cells, including T cells, B cells and natural killer (NK) cells, and helps regulate inflammatory responses. Selenium is an essential mineral that plays a vital role in the conversion of T4 to the more active form, T3, in the body. It also helps protect the thyroid gland from oxidative damage. In its antioxidant capacity, selenium supports mitochondrial function and it helps protect against oxidative stress and inflammation in the brain, promoting neuroplasticity and synaptic function and supports the production of antioxidant enzymes.

Selenium Food Sources

- Brazil nuts
- Fish
- Shellfish
- Poultry
- Eggs
- Sunflower seeds

Vitamin A

Vitamin A plays a vital role in supporting the immune system and promoting immune function. It helps regulate the production and activity of immune cells, including T cells, B cells and natural killer (NK) cells and enhances the body's defence mechanisms against infections, viruses and pathogens. Vitamin A also helps maintain the integrity of mucous membranes, which act as a physical barrier against pathogens entering the body. Vitamin A is also essential for thyroid hormone synthesis and regulation. It helps support the production of thyroid-stimulating hormone (TSH) and maintains healthy thyroid function.

Vitamin A Food Sources

Vitamin C

As an antioxidant, vitamin C helps protect mitochondria and the brain from oxidative stress and damage; it also helps support immune function by enhancing the production and function of white blood cells, including lymphocytes and phagocytes, which are essential for fighting infections. Vitamin C supports the adrenals, as it is required to produce stress hormones helping to maintain balance when under stress; even in low lying chronic stress, vitamin C can become depleted. It is not stored in the body so we have to keep intake up.

Vitamin C Food Sources

Vitamin D

Vitamin D is involved in hormone regulation and plays a crucial role in supporting the function of the endocrine system. It helps regulate the production and activity of various hormones. Vitamin D plays a role in thyroid hormone synthesis and regulation. It also plays a crucial role in regulating immune responses and modulating inflammation and may help reduce the risk of autoimmune disorders

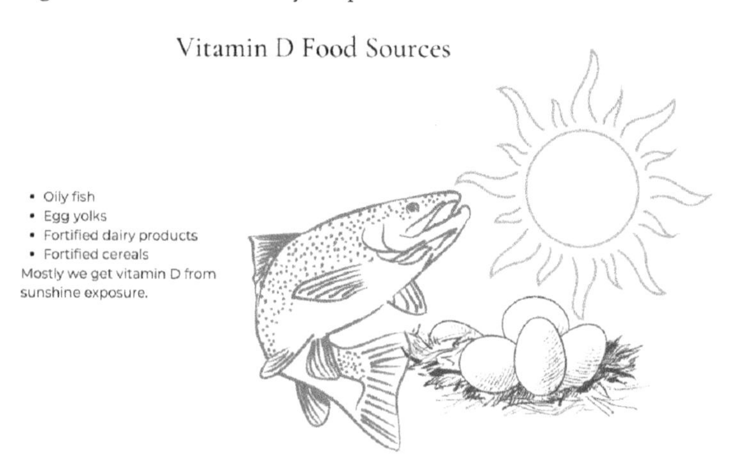

such as multiple sclerosis and Hashimoto's thyroiditis. It helps support the function of immune cells and enhances the body's defence against pathogens. Vitamin D receptors are also found throughout the brain, suggesting a role for vitamin D in brain health. Adequate vitamin D levels may help support cognitive function and reduce the risk of neurodegenerative diseases.

Vitamin E

As an antioxidant, vitamin E helps protect mitochondria, immune cells and the brain from oxidative stress and damage, and supports immune function. It helps stimulate the production and activity of immune cells, including T cells, B cells and natural killer (NK) cells and enhances the body's defence mechanisms against infections, viruses and pathogens. It also supports the adrenals by supporting hormone production and recovery from stress. Vitamin E may play a role in supporting cognitive function and reducing the risk of cognitive decline and neurodegenerative diseases. Vitamin E exhibits anti-inflammatory properties and may help reduce inflammation in the body. It helps inhibit the production of pro-inflammatory cytokines and mediators, reduce the activation of inflammatory pathways and modulate immune responses.

Vitamin E Food Sources

Zinc

Zinc is essential for the development and function of immune cells, including T cells, B cells and natural killer (NK) cells. It helps regulate immune responses, promotes the production of antibodies and supports wound healing. Zinc helps support the function of the pituitary gland, which regulates the production of several hormones; it is involved in adrenal hormone synthesis and helps regulate cortisol levels in the body. It is also involved in thyroid hormone production and regulation. It supports the synthesis of thyroid hormones and helps maintain thyroid function. Zinc plays a role in maintaining the integrity of the gastrointestinal tract. It helps support the structure and function of gut epithelial cells and may help reduce intestinal permeability.

Zinc Food Sources

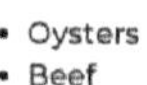

A word on supplements[5]

It's important to recognize with supplements that while they can be hugely beneficial, they may also not be right for everyone. It's easy to think supplements do no harm because they are natural, but certain health conditions, medications or other combinations of supplements and nutraceuticals will mean you shouldn't be taking a supplement. I've had many a time where someone has come to me in clinic and we have had to remove a supplement that they maybe shouldn't be on because of an underlying health concern. People read blanket information online and start popping supplements. As a BANT (British Association for Nutrition and Lifestyle Medicine) registered nutritional therapist, I not only had to study modules on supplementation, I also had to work clinically supervised in a training clinic first, showing how I made the safety and interaction checks and I also had to pass a 'safety' viva (a type of oral examination in front of a panel of examiners). This was a rigorous process and one I am very grateful to ION (Institute for Optimum Nutrition) for. Blanket recommendations and dosages are never a good idea. This being said, we have seen how some specific nutrients can benefit different pillars of our health, so supplements play a role as we can get higher strengths of these into

[5] A note on supplements: head to the online book resources section to get discounts on supplement brands we love to work with.

our system. Quality of supplements is also important and I partner with brands who take a high stance on this.

When used correctly, supplements, or nutraceuticals, can be hugely beneficial and therapeutic in supporting your health journey. For example, if we find parasitic or bacterial overgrowths, or a mould toxicity, supplemental support is usually required. If we find an imbalance in a nutrient we can use loading doses to help elevate it quickly so that we can work on getting that nutrient into the diet. Remember it's better to get as much as you can from a food source. A pill cannot write off a poor diet; they should be used as a 'supplement' to a healthy diet – yes, again the pun is intended. Here's the thing in clinic, as practitioners we are working to always stay on top of the latest research, we review brands and their quality, we look at clinical trials, no we are not using Dr Google to find a quick fix and sometimes we are testing to find out what you as an individual need. We look at dosages and forms that would be best for you as an individual and do all the safety checks.

Now I said here that we should as much as possible get our nutrients from our food, and this is always the first port of call, but we do have a problem with this in our modern society and that is soil health. We are going to look at this more in chapter 17, but essentially the way we farm, the industrialization of our modern world and the use of things like pesticides have damaged soil health and reduced the nutrients in our foods.

Here I'm going to talk you through some of the foundational supplements I use in clinic; for dosages and a supplement plan you can get in touch to work with us in clinic, and for brand recommendations and discounts where available visit the online book resources. I always recommend that you speak to a practitioner before taking supplements. I also use more advanced supplementation plans in clinic when working with certain health conditions, or underlying root causes like infections, mould, viral load and gut health. These would be tailored to the individual.

Vitamin D

Given that here in the UK we get very little sunshine, supplementation is often required. Research has shown that some health conditions such as multiple sclerosis are more prevalent in countries further away from the equator, i.e. in countries with less sunshine. Another problem, especially in areas with less sunshine, is that we only actually get enough vitamin D from the sun at certain times; in the UK, for example, it's summer between the months of April and September – this is when the sun is strong enough. You also need to have a significant amount of your skin exposed to the sun without sun cream – your arms, legs and face is ideal, without sunglasses. Obviously this is only for a limited time up to around 20 minutes, after which time we would recommend sun cream to protect you. Finally, one way to tell if you are getting vitamin

D is if your shadow is shorter than you – this is indicative that you will be able to produce vitamin D; if your shadow is longer than you, even in the summer months, it is unlikely you will be able to produce enough vitamin D. So as you can see, it's not quite as simple as you think, especially in areas with less sun exposure and we've seen all the benefits of vitamin D earlier. If you want to supplement with vitamin D you can get your levels tested. Roughly 60% of the UK population are deficient, and even 41.6% of the US population were found to be deficient, despite some areas of the US having plenty of adequate sunshine – I have worked with people in California who we have found to have vitamin D deficiency. With certain conditions like autoimmune disease it's even more important to have optimal levels of vitamin D, not just 'normal' levels. I like Designs for Health as a brand for vitamin D.

Probiotics

My go to probiotic is Symprove. We've seen the benefits of probiotics and Symprove is one I both use myself and have used with many patients. It is a water-based formula, meaning it doesn't trigger digestion so the stomach acid is not activated and more of the bacteria gets where it needs to go. They are also backed by independent science led trials, don't contain any nasties including being gluten free, dairy free and vegan, and contain 10 billion units of the beneficial bacteria. I've been using Symprove for years now and really notice a difference. I use other probiotic brands when looking to support a particular bacteria, symptom or condition.

Omega 3

This is another supplement that I use as a foundational step; we've seen the benefits earlier, but it is even more important if you struggle to get enough oily fish into your diet, or you are vegetarian or vegan. I use a brand called Bare Biology – personally I take their Life & Soul version. I also sometimes use their Mindful omega 3 which favours more DHA to support brain health. They also do a great vegan vitamin D which comes from algae, instead of fish, called Vim & Vigour. As a brand I like them because of their purity, they are well certified for both sustainability and the highest rating for being free of contaminants, which is something you need to watch with omega-3 supplements. They also use clinical dosages based on science in the literature.

Adaptogens

Adaptogens have been used for centuries, especially in Eastern medicine and Ayurvedic medicine. They are beneficial, regardless of whether you are high or low in cortisol – they work for both and they may help to build stress resilience. They are active ingredients from herbs, roots and plants that are thought to help us manage stress, or restore balance, after coming up against stress. They can help with anxiety, fatigue and overall wellbeing. Adaptogens can be really helpful, but it is also important to work on the root cause of the stress with lifestyle factors too (see chapter 16). Let's look at some of the more common adaptogens:

- Ashwagandha: may be beneficial for the endocrine, nervous, immune and cardiovascular systems and calm how the brain responds to stress. Also an antioxidant and may have anti-inflammatory properties.
- Panax Ginseng: may be supportive for mental and physical fatigue, can boost energy levels depleted due to stress and improve performance in stressful situations.
- Rhodiola: may be beneficial for fatigue, anxiety and depression and may improve performance in a stressful situation or during physical activity.

There are many adaptogens out there and sometimes with certain health conditions like Hashimoto's we need to break it down and find a mix that is safe to use, but for a general mix of stress support I like Nu Mind and Wellness (Appendix II). I use these with my optimal health clients who are busy people and need ease, or who travel and want to easily be able to pack their supplements. What is great about Nu, other than their formulations, is that they package your daily supplements into one sachet; you have a great stress formula with B vitamins, adaptogens, magnesium and more.

Glutathione/NAC

We've already seen that glutathione is your master antioxidant and NAC is a pre-cursor, so if someone struggles to take glutathione we can give them NAC instead. Quicksilver Scientific is a brand I generally love for lots of things, especially when designing an advanced supplement plan. I use them for toxic load in patients following testing. They use liposomal highly bioavailable forms and are backed by third party science. Glutathione as a supplement can commonly come in poor forms, so it's important to choose wisely.

Vitamin C

We've looked at the benefits of vitamin C for immune and mitochondrial support as an antioxidant. Some forms are better than others (ascorbic acid) and you need a good dosage for it to be effective. I like Bodybio liposomal Vitamin C as it is pure and non GMO (see chapter 17) as well as being liposomal for better bioavailability.

PC

We looked at PC under phospholipids, the building blocks of your cell membranes, repairing and protecting your cells, benefitting every cell in your body. PC may help support your brain health, cognitive performance, cellular energy and healthspan. I like Bodybio again here, as they are focused on cellular wellness which I love about them and their PC is liposomal in format, making it better absorbed.

16

Lifestyle – the 4 S's

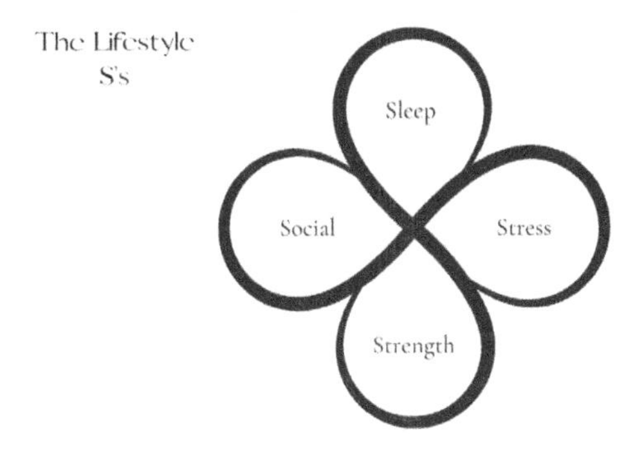

Here is what I call my 4 Lifestyle S's:

- ✓ SLEEP
- ✓ STRESS
- ✓ STRENGTH
- ✓ SOCIAL

These are proven to benefit your life, health, longevity and healthspan. They also give you more joy and happiness, which is what it is all about. Nutrition is key, but lifestyle medicine is also a huge part of what I do and how I trained. Here we will go through each S and I will give you techniques to simply integrate into your daily routine. We also have guides in the online book resources to help support you.

Sleep

PEARL: 79% of Britains struggle to switch off at bedtime and 60% wake in the night with worries.

These statistics are pretty shocking. It's thought we now sleep one to two hours less every night than our grandparents (60 years ago) – this is from studies carried out by Oxford, Cambridge, Harvard, Manchester and Surrey universities. It doesn't sound like much, but given that sleep is ideally eight hours a night, losing two hours is losing 25% of your sleep. It's also 14 hours of sleep per week, which is more than one and a half night's worth.

The stages of sleep

Understanding the stages of sleep is crucial for grasping the complexities of our nightly rest and its impact on our overall health and wellbeing. Sleep is typically divided into two broad categories: non-rapid eye movement (NREM) sleep and rapid eye movement (REM) sleep. These categories encompass the various stages of sleep, each characterized by specific brainwave patterns, physiological changes and associated experiences. Throughout the night you progress through multiple cycles of NREM and REM sleep, with each cycle lasting approximately 90 to 120 minutes. The duration of each stage within a cycle may vary, with deeper stages predominating earlier in the night and REM sleep becoming more prominent in the latter half.

Sleep Cycle

Transition to Sleep

- Onset of drowsiness
- Slowing of brain activity
- Potential for muscle contractions, or sensations of falling
- A few minutes.

Light Sleep

- 45-55% of our sleep time
- Brainwave activity continues to slow
- Bursts of brainwave activity to consolidate memory
- Muscles relax, heart rate & temperature decrease.

Deep Sleep (Slow Wave Sleep)

- Physical & mental restoration, further memory consolidation
- Repair, growth & immune function
- Blood pressure drops, muscles relax deeply, breathing slows.

Dreaming Sleep (REM Sleep)

- Unique due to rapid eye movements, heightened brain activity, vivid dreams
- Muscles paralyzed temporarily
- Emotional processing, memory consolidation & learning.

The circadian rhythm

The circadian rhythm – the body's approximately 24-hour, biological internal clock – plays a crucial role in regulating various physiological processes, including sleep–wake cycles, hormone secretion, metabolism and cognitive function. This internal clock is impacted primarily by external cues, most notably the light–dark cycle, but also temperature, social interactions and feeding patterns. At the core of this intricate system lies the suprachiasmatic nucleus (SCN), a small cluster of cells located in the hypothalamus region of the brain.

Light serves to reset your internal clock to align with the external day–night cycle. Photoreceptors in the retina detect changes in light intensity and transmit this information to the SCN, which then adjusts the timing of biological processes accordingly.

Natural light vs blue light

Natural light and blue light both play significant roles in regulating the body's sleep–wake cycle, yet they have distinct effects on sleep patterns and overall wellbeing. Natural light, particularly sunlight, serves as the primary cue for the body's internal clock, or circadian rhythm, helping to synchronize various physiological processes with the day–night cycle. Exposure to natural light during the daytime promotes alertness, cognitive function and mood while signalling the body to suppress the production of melatonin, a hormone that promotes sleep. Conversely, exposure to blue light, which is emitted by electronic devices, energy-efficient lighting and screens, particularly in the evening hours, can disrupt the body's natural circadian rhythm and interfere with sleep onset and quality. Blue light exposure suppresses melatonin production, tricking the brain into thinking it's still daytime and delaying the onset of sleep. Prolonged exposure to blue light at night can disrupt the production of other hormones, such as cortisol and growth hormone, leading to imbalances that may impact overall health.

The benefits of a good nights sleep

Mitochondrial health:

- During sleep, mitochondria undergo essential processes for energy production; adequate sleep ensures that mitochondria can efficiently carry out processes.

- Sleep plays a crucial role in promoting mitochondrial biogenesis (chapter 6).
- During sleep, cells engage in autophagy, a cellular recycling process that removes damaged or dysfunctional organelles. This helps to eliminate defective mitochondria and maintain a healthy mitochondrial population within cells.
- Sleep supports DNA repair mechanisms, reducing the risk of cellular damage and mutation.
- During sleep, cells activate antioxidant defence mechanisms to counteract ROS and minimize oxidative damage, promoting mitochondrial health and function.

Immune function:

- During sleep, the body regulates the production and activity of various immune cells. Sleep also promotes a balanced immune system by influencing cytokine release. Disrupted sleep patterns can alter immune cell populations and impair immune function.
- Sleep plays a critical role in modulating the inflammatory response; sleep deprivation is linked to increased levels.
- During sleep, the immune system ramps up its production of antimicrobial peptides and proteins, which play a crucial role in combating infections.
- Sleep is essential for the formation of immune memory, the process by which the immune system remembers past infections and mounts a more rapid and robust response upon re-exposure to pathogens.
- Sleep is essential for promoting tissue repair and wound healing.

Brain health:

- Sleep facilitates memory consolidation and learning.
- It supports cognitive function, including problem-solving and decision-making abilities.
- During sleep the brain detoxifies, removing waste products accumulated during wakefulness.
- Sleep enhances resilience, reducing the risk of mood disorders such as depression and anxiety.
- During sleep we get production of mood balancing neurotransmitters, including serotonin and dopamine.
- Sleep is essential for promoting neuronal health and protecting against neurodegeneration.

Adrenal health:

- During sleep, cortisol levels reduce, promoting relaxation.
- Sleep enhances resilience to stressors, improving overall stress management and coping abilities.

- Chronic sleep deprivation or poor sleep quality can activate the stress response, leading to increased levels of stress hormones such as cortisol.

Gut health:

- Sleep aids in digestion and nutrient absorption, optimizing gastrointestinal health.
- It supports gut microbiome balance.

Thyroid health:

- Adequate sleep is essential for the regulation of thyroid hormones, including T4 and T3. During sleep, the body undergoes various hormonal fluctuations and metabolic processes that help maintain the balance of thyroid hormones.

Hormone health:

- During sleep, various hormones are produced and secreted that help regulate hormones.
- Sleep reduces the risk of insulin resistance and diabetes by balancing blood sugar.
- Sleep supports the clearance of toxins and metabolic waste products accumulated during daily activities.
- It also enhances lymphatic system function, promoting detoxification and immune health.
- Bone health is supported by promoting the release of growth hormone, essential for bone remodelling and density.

Cardiometabolic health:

- Sleep is associated with lower blood pressure, reducing risk of hypertension and CVDs.
- Sleep supports heart rate variability, a marker of cardiovascular health and resilience.
- The regulating of appetite hormones, such as leptin and ghrelin, promotes healthy eating habits and weight management.
- Sleep plays a crucial role in regulating metabolic processes, including glucose metabolism, lipid metabolism and insulin sensitivity.

PEARL: A study showed that just one night of lost sleep can age our brain.

So how do you know if you are sleeping well?

We can look at things like how long it takes you to fall asleep, whether you sleep through the night, how many hours you get and whether you wake feeling refreshed.

Outside of this I recommend using some tech, things like the Oura ring or the Whoop band can be hugely beneficial (Appendix II). Understanding your sleep cycle and quality can aid you on your optimal health journey. In fact those who join me on my 12-month health membership are gifted technology like this to help us work on your journey together.

Sleep routine

Another thing I advise all my patients to do is to implement a sleep routine. In the online book resources you will find handouts to help you plan and implement your own personal sleep routine; if you struggle at all with sleep I strongly recommend you download these workbooks. But what does a sleep routine look like and how do we implement one? I advise my patients to work backwards. We want to aim to go to bed and get up at the same time every day so look at your week and take the earliest morning start you have. Let's say for work you need to be up at 7am, that's your wake time and you are going to have that as your wake time every day of the week. Now working backwards we want eight hours sleep, so you need to be asleep by 11pm. This is where we start to personalize it and the workbooks will help you with this. In fact go and print them off now while you are reading this section and fill them in as we go. In a matter of ten minutes you will have your own personalized plan mapped out.

Ok so how long does it take you to fall asleep? For some this might be five minutes, for others of you it could be an hour. Let's say it takes you 30 minutes. You need to be in bed then at 10.30pm to be asleep by 11pm to be awake at 7am and get 8 hours sleep. If it takes you five minutes to get to sleep you need to be in bed at 10.55pm; if it takes an hour, 10pm. You get the idea.

Your next step is to give yourself at least an hour, ideally 90 minutes, without screens or anything that will overly stimulate your brain. Instead of being on your phone, computer or TV, try reading, meditating, journalling, having a bath, having a herbal tea, listening to a podcast, whatever you find relaxing, but give yourself that time. You can also use this time as self-care wind down time, so maybe you want to get ready for bed, put on a face mask, wash your hair, whatever it might be, this time can be used to get yourself ready for bed. If you need to be in bed for 10.30pm you will start this at the latest by 9.30pm.

This really is just about making it a new habit. Habits can take time to form; it's much easier and quicker to break a habit than make one.

Morning routine

So we've talked about the evening routine, but often people forget the morning routine. This is equally important. The biggest thing you can do to help support your sleep cycle in a morning is get out into natural daylight, even if it's just for ten minutes. First thing in the morning, go and stand in the garden with the dog, or a cup of tea, walk down the road, pop to the local shop, walk to get a coffee, just sit outside, ten minutes to let natural light into your eyes. Now I know in the winter in England when it's pouring down this doesn't sound too appealing and to be honest if you haven't got a dog to make you do this you have two choices: get a dog (I highly recommend this option) or sit inside by a window looking out – even without actually being outside this will still help.

Other steps

- Set the right environment for sleep: a dark, cool room (around 17 degrees is the optimum temperature), this likely means cracking open a window.
- If necessary use black out blinds.
- Eat dinner earlier, try to allow your digestion to settle by eating dinner no later than 6pm, or 7pm if your day requires it.
- Remove screens from your bedroom – try not to have a TV in your room, I sleep much better without one. Charge your phone in a different room, or at least put it on 'do not disturb' and night mode. Ideally we would leave our phones in another room, but I have patients for whom this just doesn't work, usually when they have older teenage children, or children at university and they want to know that if their child needs them that phone will wake them. This I completely understand. To help:
    - Avoid having blue charger lights on in the room.
    - Put your phone on the night light mode to reduce blue light.
    - Keep it out of reach of your bed so you would hear it, but you can't reach it from the comfort of your bed and start flicking through Instagram!
- Stop working earlier in the evening. These days we can take our work everywhere with us. While this is highly convenient, it's also impacting our lives. Think back, our grandparents finished work at 5.30pm, clocked off and went home. They couldn't then work, they didn't have access to it, there were no laptops or emails to check, they were switched off, 5.30pm onwards they were relaxing. Try to give yourself the same self-care.
- Think about how you use your evening. I had a patient who was talking late at night to her boyfriend on the phone (the relationship was difficult at the time), another patient was watching TV but loved thrillers and horrors, another was talking to friends which was fun, but also hyping her up. All of these things, whilst they are not work, are stimulating for your brain. Try to think of things you can do instead that calm your mind. That's not to say you shouldn't talk to your friends, just do it a little earlier on and safeguard that hour before bedtime for yourself.

- Move exercise earlier in the day; exercise has stimulating effects and while we want the benefits of exercise, moving it earlier in the day can prevent it from impacting our sleep cycle.
- Move your caffeine earlier in the day, drink it pre-noon.

Sleep

CHECKLIST

- PLAN YOUR SLEEP ROUTINE
- IMPLEMENT A MORNING ROUTINE
- GET NATURAL DAYLIGHT FIRST THING IN THE MORNING
- KEEP CAFFEINE TO BEFORE LUNCH
- SCREEN FREE TIME 60–120MINS BEFORE BED
- EAT AND MOVE EARLIER IN THE DAY
- WEAR BLUE LIGHT GLASSES
- USE NIGHT MODE ON SCREENS
- CREATE THE RIGHT ENVIRONMENT
- MAKE DINNER EARLIER
- REMOVE SCREENS FROM BEDROOM
- FINISH WORK EARLIER
- EXERCISE EARLIER

The siesta

I often get asked whether it's ok to take naps. Here's the thing, I grew up part of the time in Mallorca, it's my second home and I love their way of life. Part of that is the siesta and trust me, it's a non-negotiable for them. They may not all sleep in this time, but they do have downtime. The siesta is a short nap taken in the early afternoon and it has been associated with several health benefits. Firstly it can help alleviate feelings of sleepiness and fatigue – in regions with warm climates midday heat may contribute to drowsiness. By providing an opportunity to rest and recharge, the siesta can improve cognitive function, alertness and mood, enhancing overall productivity and wellbeing. Taking a siesta has been linked to cardiovascular health benefits, including reduced blood pressure and lower risk of heart disease. So should you power nap? The Institute for Functional Medicine reported that a 10–20-minute nap reduces sleepiness; improves cognitive performance; increases alertness, attention and energy levels; improves mood; improves motor performance and reduces stress levels. A nap of 20–30 minutes enhances creativity and sharpens memory; 30–60 minutes sharpens decision-making skills, including memorization and recall and improves memory preservation; 60–90 minutes means you enter rapid eye movement (REM) sleep, which is critical for problem-solving, helps make new connections in the brain, enhances creativity, reduces negative reactivity and promotes happiness. Give it a go, see how it impacts your day; it may be something that works for you, it may not.

Stress

We've looked at the impact of stress on the body in chapter 9, but what can we do to manage stress?

Breath work

One of my favourite stress relieving techniques is breath work – it moves your body and mind from the stress response to a place of calm. There are many of these that you can do, but I like 3-4-5 breathing. It's simple, breathe in for a count of three, hold for a count of four and out for a count of five. Do ten breaths at a time. I like to do them first thing in the morning and last thing at night, and doing a set prior to eating will not only benefit stress, but also support digestion. Breathe deep into your abdomen and when you breathe out push all the air out of your lungs. Inhale through your nose and out through your mouth. If it helps you can start this practice placing one hand on your chest and one on your abdomen to help you feel your breath. As you become more adept at abdominal breathing you will find this isn't necessary. You will find some of my favourite books on breath work in Appendix II. I am particularly fond of Breathe In Breathe Out by Stuart Sandeman – he also does lovely breathwork sessions on his instagram @breathpod.

Meditation

This is one of those things that some of my patients love and some hate; you mention it and they either really want to give it a go or they are very reluctant. The truth is there are many different forms of meditation and you don't have to be sat cross legged on the floor chanting, unless you want to. For me meditation apps are the way forward – this is how I make it work for me. Jump to the online book resources to find my recommendations. You may find that a progressive muscle relaxation works best – this is where you tense and then release one set of muscles at a time, working your way up your body, i.e. your toes, your feet, your calves, your knees, your thighs, etc. Take your time and move up the body slowly. Another method is a body scan: focus your mind onto one part of your body, how does it feel? If it feels good move up the body to the next area. If you notice any tension, pain or discomfort, remain focused on that area and try deep breathing until you notice a relaxation in that part of the body, then continue on with your scan.

If the body work isn't for you, you may want to try slow deep breaths while naming five things you can see, five you can hear, five you can touch, five you can smell and if you can, five things you can taste. There are many methods and often it comes down to finding the one that works for you.

PEARL: For both breath work and mediation it's important to remember that you aren't trying to quieten your mind. People often see it as a fight; instead what you are looking for is acceptance. Feel the uncomfortable thought pass though your mind, acknowledge it and then let it go.

Gratitude

Gratitude journalling is something that has become more common in recent years. It's simple: before bed write down three things you are grateful for that day. Some people like to simply do this as a list, others prefer free writing for 15 minutes. If you don't want to do this every night, aim to do it twice a week. If you are new to this practice start with simple things: a place to sleep, your family and then as you get into the practice try focusing on the smaller things, little daily acts of kindness, or simple things that occurred that made you happy. You can further your practice by trying not to repeat things you have previously written, this makes it more challenging and forces you to lean more into your gratitude.

Journalling

While gratitude is a form of journalling you may prefer free form journalling, or using a mindfulness journal that has guided prompts in it. This can be a very soothing technique that allows you time in your day to decompress. What you write and how much you write is entirely up to you, especially with the free form style. Journalling gives you brain space; you are taking all the thoughts from your day out of your head and getting them onto paper, freeing up your mind. It can also be really useful in decision-making processes; writing down our thoughts is like sharing them with a friend, it can help you to process them. You may want to keep a journal by your bed and have this as part of your bedtime wind down routine. You can download The Goode Health Journal from the online book resources to free form write or use the guided prompts.

Relaxing hobby

Use a relaxing hobby you already have, or start a new one as a way to decompress – this could be any number of things. I have patients who play a musical instrument, listen to music, paint, draw, do some gardening; for me it's always been reading. Carve out 30 minutes a day to practice this hobby without interruptions – again this may be part of your evening routine.

Support

Lastly if you need support with mental health always reach out to a qualified practitioner to get support.

Stress

CHECKLIST

- 10 X 3-4-5 BREATHS FIRST THING IN THE MORNING AND LAST THING AT NIGHT (PLUS BEFORE MEALS IF YOU WISH)
- CHOOSE A FORM OF MEDITATION TO TRY AND COMMIT TO **10MINS** A DAY
- COMMIT TO A FORM OF GRATITUDE OR JOURNALLING AND ADD 15MINS INTO YOUR EVENING ROUTINE FOR IT
- TAKE UP, OR PRACTICE, A RELAXING HOBBY FOR 30MINS EVERY DAY
- SEEK SUPPORT IF YOU NEED ADDITIONAL HELP

Remember the key is to use these techniques all the time, not just when we are stressed. By practicing them when you are calmer, it will be easier for you to implement them when you are stressed. If you just try to newly implement it while in a heightened state, the chances are you will find it too difficult. You need to learn these tools so that when you need them you can pull one out of the bag.

Strength

Your body was built to move. It's that simple, and let's be honest, we all know it. Your risk of chronic disease and a lowered healthspan is significant if you don't move your body.

The benefits of exercise on the body and mind:

- Improved brain function
- Improved mood
- Better learning
- Improved memory
- Elevated cognitive function
- Better brain ageing
- Better mitochondrial function
- More energy
- Improved circulation
- Lower blood pressure
- Lower cholesterol
- Improved blood sugar balance
- Balanced cortisol levels
- Balanced hormones
- Improved cardiovascular risk
- Lowered inflammation
- Improved immune function
- Improved detoxification
- More diverse gut bacteria
- Healthier gut function
- Improved muscle mass and strength
- Better flexibility and movement
- Increased mobility with age

Over the years and decades we've seen various fitness waves sweep the globe, from the 80s keep fit videos to the HIIT workouts of today. The truth is, you don't need to be doing intense exercise. If you want to and it works for you, fine, but there are other options and this is why this section isn't called exercise. I usually refer to it as building strength, or movement, with my patients. I don't do any high intensity workouts, neither do many of the people I work with; it's not for everyone, and your underlying health may impact this. If we look at the blue zones work[6] by Dan Buettener we see movement incorporated into daily life and this is how I like to see it work. Those who stayed active, had physical jobs, walked around their towns, did the gardening and household chores were those who extended their healthspan.

Building strength in your body and activating your muscles, using them so you don't lose them, is key. Finding something that works for you is also the way to stay consistent.

The love of a sport

Maybe you have a sport that you love, football, cricket, or for me tennis. It may be that you normally watch this sport and you follow along with the pros, but why not get out there and play. Find a team, get some friends together, go have fun.

Yoga

Yoga offers a myriad of health benefits that extend beyond physical fitness, encompassing mental, emotional and spiritual wellbeing. Physically, yoga promotes flexibility, strength and balance through a combination of poses (asanas) and flowing movements. Regular practice can improve posture, alleviate muscle tension and enhance joint mobility, reducing the risk of injury and promoting overall physical resilience and improved healthspan. Yoga is also known for its stress-relieving properties, incorporating breathwork (pranayama) and mindfulness techniques that help calm the nervous system, reduce cortisol levels and promote relaxation. Yoga can cover your Strength and Stress pillars of lifestyle medicine.

Pilates

Pilates is a highly effective form of exercise that offers numerous health benefits, focusing on core strength, flexibility and overall body alignment. Pilates is hugely beneficial to our healthspan, improving balance, posture, mobility and core strength. This can all help keep you independent and active for longer, as well as reducing joint pain and stiffness as we age. Through a series of controlled movements and

[6] The blue zones are areas that have been identified to have more centenarians than anywhere else in the world. Dan Buettener is a researcher on longevity and blue zones. See Appendix II.

mindful breathing techniques, Pilates works to strengthen the deep stabilizing muscles of the abdomen, pelvis and back, improving posture and spinal alignment. Pilates offer the same mind–body connection for relaxation and stress relief as yoga. Pilates can be adapted to suit individuals of all fitness levels and abilities. Plus the focus on core strength and stability is why it's often my first recommendation to patients with chronic illnesses or autoimmune diseases, even chronic fatigue.

Walking

Walking

CHECKLIST

- ✓ INCREASE YOUR STEPS
- ✓ A LEISURELY STROLL AROUND YOUR VILLAGE
- ✓ A BRISK LUNCHTIME WALK
- ✓ A SCENIC HIKE
- ✓ AND IF YOU DON'T LIVE IN THE COUNTRYSIDE WHY NOT SPEND A SUNDAY HAVING AN ENJOYABLE DRIVE OUT TO THE COUNTRYSIDE FOLLOWED BY A WALK AND A NICE SUNDAY ROAST AT THE PUB, INCORPORATING **STRESS** MANAGEMENT, **STRENGTH** BUILDING AND **SOCIAL** CONNECTION, PLUS ALL THE FRESH AIR WILL LIKELY HELP YOU **SLEEP** AND DAYLIGHT BALANCE YOUR CIRCADIAN RHYTHM. **ALL 4 LIFESTYLE S'S IN ONE.**

Walking is a simple yet powerful form of exercise that offers numerous health benefits for people of all ages and fitness levels. It is also accessible to anyone at no cost. As a weight-bearing activity, walking helps strengthen bones and muscles, improve joint health and increase overall cardiovascular fitness. The key is consistency. Regular walking has been shown to lower blood pressure, reduce cholesterol levels and decrease the risk of heart disease, stroke and type 2 diabetes. Walking is an effective way to manage weight and support healthy metabolism when combined with a balanced diet. Beyond its physical benefits, walking is also beneficial for mental and emotional wellbeing. One of the best parts of walking is spending time outdoors in nature, which can reduce stress, boost mood and enhance feelings of relaxation and wellbeing. Walking provides an opportunity for reflection, mindfulness and connection with oneself and the surrounding environment, promoting emotional resilience. It's also a great time to listen to a podcast and tie in one of your hobbies.

A note on elevation

In blue zone communities where residents have lived in hilly regions and engaged in regular uphill and downhill walking throughout their lives, find the impact on health and longevity is profound. Uphill walking challenges the cardiovascular system, increasing heart rate and oxygen consumption, while downhill walking engages muscles in the lower body, improving strength, balance and coordination. If you are building your walking up start slow, but if you are ready to challenge yourself a bit more, think about elevation.

Simple steps

For more on simple steps have a look at Appendix II for my favourite books, but I always point patients to Dr Rangan Chatterjee's first book, The Four Pillar Plan. His chapter, Move, is a perfect example of how exercise doesn't need to be intense or a struggle, it can be fitted into our daily lives to suit any level of fitness. I was in fact lucky enough to meet Rangan at the launch of this book and still have my signed copy in which he wrote me a good luck message for my career. Now here I am writing my book. Doctors who are implementing lifestyle medicine into their practices is a sign we are, as a community, heading in the right direction. I'm seeing more and more of a shift this way and it fills me with positivity for the healthcare of the future. Why not try implementing Dr Chatterjee's five-minute kitchen workout. It's simple, but effective, and really there's no excuse as we can all spare 5 minutes.

Simple Strength

- PARK FURTHER AWAY FROM YOUR DESTINATION AND WALK THE FINAL PART
- TAKE THE STAIRS NOT THE LIFT
- DESK STRETCHES – GET UP FROM YOUR DESK EVERY 50MINS FOR 10MINS AND WALK AROUND
- SPEND 30MINS OF YOUR HOUR LUNCH BREAK GOING FOR A WALK
- IF ON A PHONE CALL STAND UP, OR WALK AROUND YOUR OFFICE, RATHER THAN BEING SEATED
- BUY A STANDING DESK OR A WALKING PAD FOR UNDER YOUR DESK

The 3 P's

Pacing–Planning–Prioritizing

One of the biggest challenges my patients face with movement is incorporating it when they are lacking energy, or struggling with fatigue.

The spoon theory is a widely recognized concept used to explain pacing and energy management for those with chronic illnesses, or fatigue conditions. Your spoons represent units of energy that are available to you to use each day; your spoons are limited much like your energy. Every activity, from getting out of bed to completing tasks throughout the day, requires a certain number of spoons. Tasks that require more energy use more spoons. Once all spoons are used up, this is the time you will experience significant fatigue. By pacing yourself you practice conserving spoons so you don't push yourself to the point of total exhaustion. So if one day you are going to a Pilates class, this may use up two spoons – on this day you may not want to include a work task that also takes up lots of spoons. The goal is for you to learn to allocate limited energy to tasks so that you prioritize and plan ahead.

Learn to prioritize tasks, what needs doing that day, and focus on those first; then if you have to push some tasks to the following days, it won't be the ones that needed doing. Some days work will need to be a priority, other days you may be able to prioritize a walk, find a balance. Also start planning your days. If you have a crazy busy work day with back to back meetings, that's probably not the day to do your workout. I like to spend an hour on a Sunday planning out the week ahead. Think about your spoons and try to balance the use of energy though the week evenly. Also schedule in your self-care as well as things for work. Put your yoga in the diary, plan in your journalling time, or allocate time for that hobby.

Pace–Plan–Prioritize.

Don't retire

Ok so you may actually want to retire from your job, but what I mean by this is don't retire, sit in a chair and do nothing. Continue to work your body with things like gardening, house work, caring for the grandchildren, cooking. Keep your body moving, continue with something you've always done like walking or Pilates, or start something new, it's never too late.

Movement can also be great for social connection, which we are going to come onto the benefits of now in our fourth Lifestyle S. Walk with friends, play a team sport, go to a yoga or Pilates class. Build social connections through the use of building strength. On that note, let's look at social connections.

Social

The fourth of my Lifestyle S's is about social connection, building a community around you, protecting beneficial relationships and banishing others, building boundaries, learning to say no, finding value and your purpose.

Build your community

Building a strong community is paramount for overall health, longevity and vitality, for many reasons. Human beings are inherently social creatures, and our connections with others profoundly impact our physical, mental and emotional health. Functional medicine emphasizes the interconnectedness of various aspects of health; community serves as a cornerstone of health, providing support, connection and a sense of belonging that are essential for optimal health and vitality.

One of the key ways in which relationships impact health is through the support and encouragement provided by positive social connections. Studies have shown that individuals with strong social support networks tend to have better physical health outcomes, lower rates of chronic disease and increased longevity compared to those who are socially isolated.

PEARL: Loneliness is the silent killer, laughter is your best medicine.

Supportive relationships provide emotional reassurance, practical assistance and a sense of security that can buffer against stress and adversity, enhancing resilience and promoting overall wellbeing. Whether it's a close-knit family, a group of friends or a supportive community, having people to lean on during challenging times can make a significant difference to one's health and quality of life. Also being able to support others in your community can provide you with a sense of purpose and meaning.

A word on toxic behaviour

Conversely, toxic or negative relationships can have detrimental effects on health and wellbeing. Toxic relationships, characterized by conflict, criticism and emotional manipulation, can contribute to chronic stress, anxiety and depression. It is well researched that the constant strain of navigating difficult relationships can weaken the immune system, disrupt hormonal balance and increase the risk of developing chronic health conditions such as heart disease, autoimmune disorders and digestive issues. Recognizing the harmful effects of toxic relationships and taking steps to remove oneself from harmful dynamics is essential for preserving health and promoting healing. Toxic relationships are not worth having in your life; one thing for sure is these people are never happy and they will only drag you down. It's best to remove yourself, it will free you from a world of pain, literally and emotionally. Instead focus your attention on those people who lift you up.

Beyond providing emotional support, belonging to a community can foster a sense of purpose and identity. Engaging in social activities, volunteering or participating in group events not only strengthens social bonds, but also promotes physical activity, cognitive stimulation and a sense of connection to something greater than oneself. Research has shown that individuals who are actively engaged in their communities tend to have better mental health, cognitive function and resilience to stress, leading to improved overall health outcomes and longevity.

Social connections have also been linked to positive health behaviours and lifestyle choices. People who are part of a supportive community are more likely to engage in regular exercise, nutritious eating and seeking preventive healthcare. Having friends or family members who encourage healthy habits and hold one accountable can increase motivation and adherence to positive behaviour changes, leading to better health outcomes in the long term.

So invest in relationships, nurture social connections and walk away from toxic people.

Learn to say NO

PEARL: NO is a complete sentence.

I remember the first time I heard someone say this, it was a lightbulb moment. No doesn't need an explanation. People-pleasers are usually the ones who end up ill. It's ok to set boundaries. Let this be your permission to set those boundaries for yourself, to say no without explanation and to put yourself first.

The importance of saying no and setting boundaries cannot be overstated when it comes to maintaining mental, emotional and physical wellbeing. Constantly saying yes to others' demands and neglecting one's own needs can lead to chronic stress, burnout and compromised health. People-pleasers, those who habitually prioritize others' needs over their own, often find themselves stretched thin, unable to establish healthy boundaries or assert their own priorities. Women tend to be more prone to doing this. This pattern of behaviour can have profound consequences on both mental and physical health, contributing to conditions such as anxiety, depression, chronic fatigue and autoimmune disorders. Continuously suppressing one's own needs and desires in favour of pleasing others can create a chronic state of internal conflict and resentment, eroding self-esteem and undermining relationships. Learning to say no and set boundaries is essential for preserving energy reserves and fostering self-care practices that promote overall wellbeing.

Your values and your purpose

Having a clear sense of purpose, or 'plan de vida', and living in alignment with one's values are essential components of holistic health and wellbeing. A sense of purpose provides you with a deep-seated motivation and direction in life, guiding your actions, decisions and priorities. It encompasses a sense of meaning and fulfilment, whether it's pursuing a passion, fulfilling a mission or making a positive impact on the world.

Research has shown that individuals with a strong sense of purpose tend to have lower rates of chronic disease, reduced risk of mortality and increased longevity compared to those who lack a clear sense of direction in life. Having a purpose promotes resilience to stress, enhances immune function and improves overall cardiovascular health. Living with purpose is associated with greater levels of life satisfaction and happiness, contributing to a higher quality of life and improved mental health outcomes.

Living in alignment with one's values is closely intertwined with having a sense of purpose. Values represent the guiding principles, beliefs and ideals that define what is important and meaningful to you, they will be different for everyone. Living in line with your values means you experience a greater sense of authenticity, integrity and coherence in your life. This alignment fosters a sense of inner peace, contentment and harmony, promoting psychological wellbeing and emotional resilience. Research has shown that individuals who prioritize their values and engage in behaviours that are consistent with their core beliefs tend to have better health outcomes, including lower levels of stress, improved immune function and reduced risk of chronic disease.

Finally, having a clear sense of purpose and living by one's values provides you with a roadmap for decision-making and goal-setting. When faced with difficult choices or life transitions, having a strong sense of purpose and clarity about what truly matters can guide you to making the right decisions for you. This sense of direction leads to greater success.

Give yourself a day off

While social connections are important and building your community is a positive thing, it's also ok to give yourself a day off. Taking a day off to enjoy solitude and quiet reflection away from relationships and social interactions is not only acceptable, but can also be beneficial for overall wellbeing. In today's fast-paced and interconnected world, it's easy to become overwhelmed by constant stimuli and demands from others. Giving yourself permission to retreat and recharge in solitude allows for introspection, self-care and replenishment. This quiet day offers an opportunity to tune into your inner thoughts and feelings, cultivate mindfulness and restore a sense of balance and harmony within.

17

Quality of our food

We've talked a lot about what to eat and nutrients, but we are still missing one really big piece of the puzzle, the quality of our food. Think about it logically, if you eat all the foods you know you need to but those foods don't have the nutrients in, you are not going to get any benefit from eating the foods, right?

Organic food

Did you know there is a difference between soil and dirt? Soil contains all the nutrients, it is fertile, dirt is not. We cannot just use and abuse our planet, we need a mutually beneficial relationship. If we use sustainable farming practices, protect our oceans and protect our forests, our planet gives back to us with nutrient-rich food, clean water and clean air. We are at one with nature – it's no coincidence we talked about the functional medicine tree in chapter 3. Think about the simplest connections we have: we breathe out carbon dioxide, plants breathe it in, they put out oxygen, which we breathe in. We are interlinked. Plants also send carbon down into the earth through their roots, they give the soil microorganisms carbon and the soil microorganisms give the plants nutrients, we eat the plants.

We need to eat what is in the soil that is transferred to the plants. This will only be transferred to the plants if the soil is healthy.

PEARL: Did you know that the chemical gas they used in the holocaust was the first of the pesticides developed for food to speed up its production?

How do you feel about eating organic food now? I'm guessing it just jumped up your priority list. Who wants to eat food sprayed with a chemical used to kill people? It makes zero sense, even if you have no knowledge of soil health or organic farming; that one fact alone should be enough to make you realize its importance. Have you ever actually watched them spraying this stuff on your food? Have you seen what they wear to do it? From masks to full protective hazmat suits, they are not going to breathe in more of this than they have to, so why on earth are you going to eat it?

When the soil and plants are sprayed, not only are those chemicals sat on our food so that we ingest them, they also kill the microbes in the soil that we need to give the food the nutrients.

Glyphosate

Glyphosate, the active ingredient in many herbicides, including the popular product Roundup, has raised significant concerns regarding its health ramifications. Studies have linked glyphosate exposure to various health issues, including disruptions in gut microbiota (they kill the microbes in the soil and the microbes in your gut), inflammation and carcinogenic effects. Glyphosate residues are commonly found in foods such as grains, vegetables and fruits, in fact corn in the US is almost always going to be sprayed with glyphosate now, as well as in animal products from animals fed glyphosate-contaminated feed. The widespread use of glyphosate in conventional agriculture has led to its presence in not only the food supply, but our water supply too, raising concerns about its long-term health effects on human health. While regulatory agencies argue that glyphosate residues in food are within safe limits, many health experts advocate for better regulation to protect human health.

What does it mean to buy organic food?

- Fewer pesticides, herbicides and fungicides sprayed on your food and the soil, instead using techniques like crop rotations to avoid you ingesting these toxins, but also increasing the nutrients in the food you eat.
- Higher animal welfare to protect the living conditions, what they are fed, and how they are reared. They are free range and cruel techniques used for certain benefits to battery farming are banned.
- Lower antibiotic use – unable to be used as standard practice only if an animal requires antibiotics.
- No genetically modified foods or feed.
- Artificial colours and preservatives are highly restricted.
- Better farming practices, like crop rotation and animal grazing rotation, that protect the soil.

Let's look at the antibiotic use a little more. It's common knowledge now that doctors have pulled back on dishing out antibiotics, quite rightly they are aware of the antibiotic resistant issue that comes with over prescribing these drugs and the impact on your gut microbiome, so they are using them when required and not before. But here is the problem:

PEARL: 30% of antibiotic use in the UK is in non-organic farming given to animals as standard practice to prevent them getting ill, not because they need the antibiotic.

So the doctors aren't dishing them out, but you are buying them from the supermarket. Remember you aren't just what you eat, you are what you eat eats.

Organic CHECKLIST

- ✓ BUY ORGANIC EVERYTHING IF YOU CAN.
- ✓ IF BUYING EVERYTHING IS TOO EXPENSIVE LOOK TO THE DIRTY DOZEN & CLEAN FIFTEEN LIST, WHICH GIVES YOU THE FOODS EACH YEAR THAT ARE CARRYING THE HIGHEST LEVELS OF CONTAMINANTS AND THE LEAST. FIND THE LIST IN APPENDIX II.
- ✓ BUY MEAT THAT IS GRASS FED.
- ✓ BUY FISH LABELLED WILD, LINE-CAUGHT, FRESHWATER OR ATLANTIC.
- ✓ EGGS THAT ARE FREE-RANGE.
- ✓ DAIRY FROM GRASS FED SOURCES.
- ✓ OILS THAT ARE EXTRA VIRGIN, UNREFINED OR COLD PRESSED.
- ✓ TRY GROWING YOUR OWN FRUIT & VEG, NOT ONLY IS IT LESS EXPENSIVE, YOU ARE IN CONTROL OF WHAT GETS ON YOUR FOOD.
- ✓ PLANT YOUR OWN HERB GARDEN.
- ✓ SUPPORT ORGANIC FARMERS WHO ARE USING SUSTAINABLE PRACTICES, I LOVE A DAYLESFORD, OR RIVERFORD, I ALSO GET THE BENEFIT OF DIFFERENT VARIETIES.
- ✓ EDUCATE YOURSELF. LEARN ABOUT OUR PLANET, FARMING PRACTICES & FOOD. ASK QUESTIONS, BE INTERESTED, BE PART OF THE SOLUTION.

Organic farming has much better conditions for the animals. Optimizing their lifestyles has the same impact on their health that it does on human health. The animals don't get sick as much and as organic farming is only allowed to use antibiotics if a sick animal really needs them, the antibiotics in your foods are significantly lowered and potentially nonexistent.

Now in the UK, genetically modified foods are relatively limited currently, although there has been some talk of this increasing. Some countries such as the US have more already on the market. However, we aren't safe from this in the UK as non-organic farming can use genetically modified feed to give the animals and as we have already said, you are what your food eats.

So what steps can you take?

Seasonal eating

Seasonal eating is rooted in the principle of consuming foods that are naturally available during specific times of the year, the idea being that it offers a multitude of benefits for both health and sustainability. This practice aligns with the rhythms of nature and our connections with nature, allowing us to nourish our bodies with fresh, nutrient-dense foods that are at their peak flavour and nutritional content. By embracing seasonal eating, we not only support local agriculture and reduce their carbon footprint, but we also optimize our health benefits by consuming foods that provide the nutrients and energy needed to thrive during each season.

Nutrient density and flavour

One of the primary advantages of seasonal eating is the superior taste and nutrient density of fresh, in-season produce. Fruits and vegetables, that are not forced and are harvested at the peak of ripeness and consumed shortly thereafter, retain their natural flavours, textures and most importantly their nutritional integrity. These foods are rich in essential vitamins, minerals, antioxidants and phytonutrients that support immune function, cellular repair and overall vitality.

Optimal nutrient profile

Seasonal eating also ensures that individuals receive the nutrients and energy needed to adapt to the changing environmental conditions of each season. Nature provides foods that are naturally suited to support the body's physiological needs during different times of the year. For example, summer fruits like watermelon and cucumber have high water content to help keep the body hydrated and cool in hot weather, while winter root vegetables like carrots and sweet potatoes are rich in carbohydrates and antioxidants to provide sustained energy and immune support during colder months. By consuming foods that align with the seasonal climate, we can optimize our nutrient intake and support our body's natural rhythms.

Supporting local agriculture

Seasonal eating promotes sustainability by supporting local farmers and reducing the environmental impact of food production and transportation. When we prioritize locally grown, or independent organic farms and seasonal produce, we support regional agriculture and contribute to the preservation of farmland and biodiversity. By reducing the distance that food travels from farm to table, seasonal eating helps minimize carbon emissions associated with transportation and storage, leading to a smaller ecological footprint. Additionally, supporting local farmers strengthens community resilience and food security, fostering connections between individuals, farmers and the land.

Farm-to-table food

The farm-to-table movement represents a return to traditional, sustainable food systems that prioritize local sourcing, seasonal ingredients and direct relationships between producers and consumers. At its core, farm-to-table dining emphasizes transparency, traceability and the celebration of local flavours and culinary traditions. It is all about knowing the provenance of your food, educating yourself, knowing where your food comes from and how it has been grown or raised. Farm-to-table food, like seasonal eating, celebrates local agriculture and the diversity of regional food landscapes. Farm-to-table food also means you will be eating seasonally and supporting local farmers.

If we want to optimize our health, increase healthspan, live longer, age better and avoid chronic illness, we have to protect the earth, the whole ecosystem of our planet from soil, air and water to bacteria, plants and animals. This is how humans thrive. ***Be part of the solution, not part of the problem***.

18

Environmental toxins

We've already talked about the impact of pesticides and herbicides on food as an environmental toxin in chapter 17, but what about other environmental toxins and their impact on our health? They are something that I come across as root causes, or as having a significant impact, on health in clinic. I could write a whole book on environmental toxins, especially given the amount of cases I've worked on with these, but here what I want to do is give you an introduction to the environmental toxins we face and the impact they have on our bodies and our health.

Mould

Mould, or mycotoxins, are one of the biggest root causes outside of stress that I come across in clinic. I've worked with cases where mycotoxins are a root cause in Hashimoto's, polyreactive autoimmunity, multiple sclerosis, chronic fatigue, psoriasis, asthma, intestinal permeability, unexplained skin rashes, joint pain, brain fog, anxiety, fatigue, sinus issues and more. Mycotoxins have also been linked to Alzheimer's, autism, infertility, inflammatory bowel disease and Parkinson's. Mycotoxins are a group of toxic compounds produced by certain fungi that commonly contaminate food crops, water-damaged buildings and indoor environments.

A word on Ochratoxin A

One of the most common mycotoxins I find in clinic is Ochratoxin A. I want to add a word on this here as it is what we call a persister mould. It binds to a protein in our blood called albumin and then not only can it not be eliminated from the body, it can also travel around the body impacting almost every body system and cause a wealth of symptoms. It is commonly found in water-damaged building and it's important to note you do not need to be able to see the mould in your home or office in order to have it in a building.

Mycotoxins can be neurotoxic, hepatotoxic, immunotoxic, nephrotoxic, GI toxic, carcinogenic, toxic to the reproductive system and cause oxidative stress. They are slowly becoming more researched and understood as they play a huge role in the development of chronic illnesses.

We can test for mycotoxins via a simple urine test (chapter 19). Much of what we do to reduce toxin load is via advanced supplementation, but there are some other things you can do to support yourself. Implement preventive measures to minimize mould growth in indoor environments. Ensure proper ventilation, maintain optimal humidity levels (ideally below 60%), promptly address water leaks and moisture problems and conduct regular inspections for signs of mould growth. Minimize mycotoxin contamination in food and beverages; store grains, nuts and other susceptible foods in airtight containers in a cool, dry place; and discard any mouldy or visibly contaminated food items. Use high-quality water filtration systems to remove mould spores and mycotoxins from drinking water. I have a quooker fitted in our kitchen, but there are other options. Use air purifiers equipped with HEPA filters to remove airborne mould spores and mycotoxins from indoor air. I like Air Doctor – visit the online book resources for a discount code. Place air purifiers in commonly used areas, such as bedrooms and living rooms, to reduce exposure to mould toxins. When cleaning mould wear protective gear to avoid touching or breathing in the mould spores. Support detoxification with antioxidant rich foods, cruciferous vegetables, garlic and high fibre foods – jump back to chapters 12 and 14 for more on nutrition for detoxification.

Heavy metals

Heavy metals are naturally occurring elements with high atomic weights and densities, including lead, mercury, arsenic and cadmium, among others. While trace amounts of these metals are necessary for normal physiological function, excessive exposure can lead to serious health problems, including impaired energy levels and compromised immune function.

Heavy metals can impair mitochondrial function, induce oxidative stress, promote inflammation, deplete antioxidant defences, weaken the immune system and increase susceptibility to infections and chronic diseases.

Heavy metals can enter the body through various routes, including contaminated water, food, air and consumer products. Drinking water contaminated with lead, mercury, arsenic or cadmium from industrial runoff, mining activities or ageing infrastructure can contribute to heavy metal exposure. Lead pipes, mercury-containing dental amalgams and contaminated groundwater are primary sources of heavy metals in drinking water. Consumption of foods contaminated with heavy metals, such as seafood (mercury), rice (arsenic) and leafy greens (cadmium), can contribute to dietary exposure. Eating organic is beneficial here. Inhalation of airborne particles containing heavy metals, such as lead, arsenic and cadmium, from industrial emissions, vehicle exhaust and combustion processes can lead to respiratory exposure.

Heavy metal toxicity can manifest with a wide range of symptoms, depending on the type of metal, the level of exposure and individual susceptibility. These symptoms may affect various organ systems and can be acute or chronic. Symptoms include fatigue and weakness; neurological symptoms such as headaches, dizziness, tremors, memory loss, cognitive impairment, neuropathy, gastrointestinal symptoms, muscle and joint pain; cardiovascular symptoms such as hypertension, palpitations and arrhythmias; skin issues; mood and behavioural changes; immune dysfunction; renal impairment; reproductive or developmental impacts.

We can test for heavy metal exposure (chapter 19) which can allow us to tailor protocols to you. We also look at supporting detoxification pathways, hydration, electrolyte balance and nutritional support.

Fish

Oily fish are full of the omega-3 fatty acids which are anti-inflammatory, however we often hear about heavy metals and pollutants in fish. Mercury in particular is found in some fish in a higher level and if it builds up it can be neurotoxic, immune toxic and toxic to the digestive system.

Fish with higher mercury levels	Fish generally considered safe with lower levels
Shark	Salmon
Ray	Sardines
Swordfish	Herring
Marlin	Trout
King Mackerel	Canned light tuna
Orange Roughy	Pollock
Southern Blue Fin Tuna	Shellfish

How fish is caught, or whether it is farmed, also impacts the toxin levels as well as nutrient density and quality (see chapter 17).

Endocrine disruptors

Endocrine disruptors are a group of chemicals that interfere with the body's endocrine system, disrupting hormonal balance and function. Among these endocrine-disrupting chemicals (EDCs), compounds like bisphenol A (BPA), phthalates and parabens have garnered significant attention for their adverse effects on human health. These chemicals mimic, or block, hormone action, leading to dysregulation of endocrine function and disruption of developmental, reproductive, neurological and immune systems. BPA, commonly found in plastics, epoxy resins and food packaging, is known for its oestrogenic activity and has been linked to hormonal imbalances, reproductive disorders and immune dysregulation. Phthalates, widely used as plasticisers in consumer products, including plastics, personal care products and building materials, can interfere with hormone signalling and have been associated with adverse reproductive outcomes, asthma, allergies and immune suppression. Parabens, commonly used as preservatives in cosmetics, skincare products and pharmaceuticals, have oestrogenic properties and can disrupt endocrine function, leading to hormonal imbalances, reproductive disorders and immune dysfunction. Common sources of exposure to endocrine disruptors include plastics, food and beverage containers, canned foods, personal care products (such as shampoos, lotions and cosmetics), household cleaners and pesticides.

Non-toxic cookware

Essential for promoting overall health and minimizing exposure to harmful chemicals. Traditional cookware materials, such as non-stick coatings (containing perfluorooctanoic acid, or PFOA), aluminium and certain types of plastics, can leach toxic substances into food during cooking, particularly when exposed to heat, or acidic ingredients. To avoid these risks, consider opting for non-toxic cookware options such as:

- Stainless Steel: Stainless steel cookware is a popular choice for its resistance to corrosion and non-reactive properties. Look for high-quality stainless steel cookware without non-stick coatings or aluminium cores.
- Cast Iron: Cast iron cookware is renowned for its excellent heat retention and natural non-stick properties. Seasoned properly, cast iron pans develop a protective layer that prevents food from sticking and eliminates the need for chemical coatings.
- Glass: Glass cookware, such as baking dishes and casserole dishes, are inert and non-reactive, making it a safe option for cooking and storing food.

- Ceramic: Ceramic cookware, made from clay and other natural materials, is free from harmful chemicals and provides even heat distribution. Look for ceramic cookware with non-toxic glazes and avoid products containing lead or cadmium.

Plastics are everywhere – our food is usually wrapped in it, water bottles are made from it, you will even find it all over our oceans now. Plastics pose significant challenges to both human health and the environment. From production to disposal, plastics release harmful chemicals and contribute to pollution. Plastic debris contaminates waterways and oceans, endangering marine life and ecosystems, while microplastics infiltrate food chains and ecosystems, potentially harming human health. Additionally, plastic waste overwhelms landfills, leading to soil and water contamination. Plastic packaging, particularly single-use containers and bottles, can release harmful chemicals such as BPA, phthalates and styrene into food and beverages, especially when exposed to heat or acidic conditions.

Air pollutants

Air pollution poses significant risks to human health, with various pollutants affecting respiratory health and immune function. Particulate matter (PM), nitrogen dioxide (NO2), sulphur dioxide (SO2) and volatile organic compounds (VOCs) are among the most common air pollutants that can have detrimental effects on both the respiratory system and immune function. Indoor air purification systems, such as high-efficiency particulate air (HEPA) filters (Appendix II), can help remove airborne pollutants and improve indoor air quality. Proper ventilation and air filtration in indoor spaces are essential for reducing exposure to indoor air pollutants and allergens. Additionally, avoiding outdoor pollution hotspots, such as high-traffic areas and industrial zones, during peak pollution hours can minimize exposure to outdoor air pollutants. Air pollution can increase respiratory irritation, exacerbating asthma and other respiratory conditions, and increasing the risk of cardiovascular diseases. They can trigger systemic inflammation and oxidative stress, compromising immune function and increasing susceptibility to infections.

Beauty products

Choosing natural, clean beauty products and cosmetics offers numerous benefits for our health and wellbeing. Conventional beauty products often contain a plethora of synthetic chemicals, fragrances, preservatives and other potentially harmful ingredients that can be absorbed through the skin and into the body. These chemicals may disrupt hormone function, irritate the skin and contribute to a range of health issues. By contrast, natural and clean beauty products are formulated with plant-based ingredients, botanical extracts and essential oils, which are gentler on the skin and less likely to cause adverse reactions. They also often contain antioxidants, vitamins and nutrients that nourish and protect the skin, promoting a radiant complexion and supporting skin health in the long term.

Chemicals you should avoid in products:

- Parabens: Commonly used as preservatives in cosmetics, skincare products and pharmaceuticals, parabens (such as methylparaben, ethylparaben, propylparaben and butylparaben) have been associated with hormone disruption and potential links to breast cancer.
- Phthalates: Phthalates are often used as plasticisers in personal care products (such as fragrances, lotions and nail polishes) and pharmaceuticals, but they can disrupt hormone function and have been linked to reproductive and developmental issues.
- Sodium Lauryl Sulfate (SLS) and Sodium Laureth Sulfate (SLES): These surfactants are commonly used in shampoos, body washes and toothpaste to create foam, but they can strip the skin of natural oils, leading to irritation and sensitivity.
- Formaldehyde and Formaldehyde-Releasing Preservatives: Found in certain personal care products, including shampoos, body washes and cosmetics, formaldehyde and its releasers (such as DMDM hydantoin, imidazolidinyl urea and quaternium-15) are known carcinogens and skin irritants.
- Triclosan and Triclocarban: These antimicrobial agents are often found in antibacterial soaps, hand sanitisers and other personal care products, but they can disrupt hormone function and contribute to antibiotic resistance.
- Polyethylene Glycols (PEGs): PEGs are commonly used in cosmetics and pharmaceuticals as thickeners, solvents and softeners, but they can be contaminated with potentially harmful compounds such as 1,4-dioxane, which is a known carcinogen.
- Artificial Fragrances and Dyes: Synthetic fragrances and dyes used in personal care products and pharmaceuticals can contain a mixture of potentially harmful chemicals, including phthalates and allergens,

which can cause skin irritation, allergic reactions and respiratory issues.

- Petroleum-Based Ingredients: Petroleum-derived ingredients, such as mineral oil, petrolatum and paraffin wax, are commonly used in skincare products, but they can clog pores, disrupt skin barrier function and may be contaminated with harmful impurities.
- Oxybenzone and Octinoxate: These chemical sunscreens are commonly found in sunscreens and can disrupt hormone function and damage coral reefs when washed off into the ocean.
- Toluene: Found in nail polish and hair dyes, toluene is a solvent that can cause neurological damage and respiratory issues with prolonged exposure.

For a list of my favourite brands head to Appendix II.

19

Functional testing

So you've made it this far. You've implemented The MitoImmune Nutrition Plan, you've taken it to the next level and looked at specific nutrients to support the pillars of your health you need to work on, you've worked on the 4 Lifestyle S's, maybe you've even started looking at the quality of your food and your environmental toxins, but what if you really want to dig into that root cause?

Maybe you are seeing the benefits to working on your health (remember benefits and results don't come with reading this book, they come from implementing the strategies in it), but you want to really start to tailor it to you as an individual, or maybe you are just ready to reach even more optimal health and performance.

At this point in your journey you are going to want to work with a functional medicine practitioner. We will look at this some more in chapter 20, but much of this testing has to be ordered through a practitioner and will require interpretation.

Testing truly is about root cause medicine and personalizing healthcare. It is also a way of seeing on paper what is going on; for some people this is hugely beneficial, seeing it in black and white can really motivate you to make changes. I see this in clinic a lot. I can review a case and health history and I can talk through the case with the patient, but with some people it's when they see what I am talking about in writing that I can see something shift in them. So many people I've worked with have commented on how fascinating functional testing is.

PEARL: "It's the best investment I've made into my health." – former patient

This was what one of my optimal health clients said to me and he is right, it is an investment that will pay you back tenfold. The testing we use is evidence based, science backed and run by approved world class private labs in the UK, Germany, US and more. Better still, the testing can be done in the comfort of your own home – most are stool, urine or saliva tests, some are blood tests but with this a phlebotomist will come to your home or office and do your blood draw, or you can visit a local centre if you prefer.

So let's look at just a few of the options I offer in clinic. Remember this is just some of the testing that we have looked at in this book – we have a vast array of functional testing we can offer, just head to the online book resources to download our testing menu.

Full blood panel

Many blood panels contain relatively limited markers, I use advanced blood screens with a wealth of markers. The benefit of this is that rather than looking at one individual marker for something, your blood can tell your story. By looking at a collection of markers we build a picture of your health and information about root causes. I use a lab that provides advanced interpretative reports, which allows us to read markers together for symptoms or conditions and then also create a comparative review for follow up.

Full thyroid panel

Getting a full thyroid panel in the UK can be difficult. Many doctors will test TSH, but getting a full panel for thyroid function and antibodies is not easy. This can lead to many people being told that their levels are normal, or that their thyroid function looks ok and nothing more needs to be done. By reviewing a full panel we can assess your thyroid function, how you are converting thyroid hormones, whether your condition is autoimmune, whether that autoimmunity is stable and more.

Comprehensive stool test

This is your gut testing. It is an invaluable tool into your health, as we have seen a healthy gut is linked to so many areas of your body. It uses cutting edge technologies. It can provide invaluable detail about your microbiome; the balance of beneficial bacteria; any parasitic, fungi or bacterial pathogens; any overgrowths; the integrity of your gut; your biochemistry and gut function; inflammation and immune function. The test is taken over three days of samples at home and gives an approximately 20 page report to provide a full picture.

Adrenal testing

With adrenal testing we can look at your cortisol curve to see if you have adequate levels of cortisol through the day. This involves a minimum of five saliva samples taken over the course of the day and looks at your cortisol awakening response (CAR) to see if you get enough of an elevation to give you energy. We also look at your DHEA levels also involved in adrenal function. From the testing we can see what stage of adrenal function, or dysfunction, you are at.

Hormone testing

If you want to advance your adrenal testing we can run a full hormone test which includes the adrenal testing just mentioned, plus we can look at your sex hormones. It helps to identify root causes of hormone imbalances in both men and women. We can look at an oxidative stress marker, melatonin, organic acids including B vitamins, glutathione, adrenaline and neuroinflammation markers. We take this assessment in urine over a 24-hour period.

Nutritional status

We can also look at your nutritional status, identifying any deficiencies allowing us to truly tailor nutrition and supplementation to reach optimal status. Nutritional deficiencies are common in sub-optimal health and chronic illnesses.

Mycotoxin testing

A urine test simple to run at home and can assess the presence of various different mycotoxins (moulds) within your body. We can identify exposure and measure levels of various different mycotoxins.

Heavy metal testing

Heavy metal testing can look at exposure to metals in the blood. It can look at levels of arsenic, cadmium, cobalt, lead, mercury, silver and strontium, as well as elements such as calcium, copper, lithium, magnesium, manganese, molybdenum, selenium and zinc.

Viral panel

We use viral testing through a lab that does gold standard viral load testing in Germany. We can assess reactivation of viruses such as Epstein Barr Virus, HHV6, cytomegalovirus, as well as looking at Lyme disease, chronic fatigue and more. This is different testing to the UK testing and only provided by the lab in Germany. Their method has the ability to test cellular immune response.

Autoantibody testing

Advanced clinical tests for immune function and dysregulation. This testing is not diagnostic, it is predictive. There are different panels we can run. We often start with a panel that looks at autoantibodies around different body systems, which can be indicative of risk of autoimmunity to these body systems. We also have panels to look at intestinal permeability, gluten sensitivity, food immune reactivity, neurological autoimmune screen, blood-brain barrier permeability, chemical immune reactivity and more.

Nutrigenomic DNA testing

In clinic we have access to various Nutrigenomic DNA testing depending on what you wish to assess. This is the ultimate in personalized medicine. We can look at how your genes are impacting your responses, root causes and imbalances. Nutrigenomics is looking at gene expression and how you may respond to various things. We can look at how you respond to foods, nutrients, what your requirements are for nutrients, inflammatory response, detoxification, oxidative stress, how you respond to exercise, what type of exercise may be best for you, how best to fuel your body and so much more.

In summary

If you are ready to jump into personalized medicine and work with a functional medicine practitioner and you would like myself or my clinic to support you in this, you will find more information in Appendix I and in the online book resources.

20

What next?

We are at an amazing point in time for healthcare. Functional medicine and its benefits have been growing in the US and now they are growing in the UK. The science we have access to now has changed the way we approach our health and allows us not only to treat disease, but to optimize our health to promote longevity.

We are also seeing terrible setbacks as our culture of fast-paced living has promoted inflammatory, fast processed foods. Our lifestyles are heightening chronic stress, we want it all, but we don't want to put the work in. We are polluting our waters, wrapping our food in plastic, spraying our food with chemicals. We have got so many things wrong, but by learning about these things, by educating ourselves, which you are already doing by reading this book, you can take steps to mitigate the impact of these things.

We have to prioritize ourselves and our health and we have to put time into doing this. Functional medicine changes aren't easy or fast – there is no magic pill; if that's what you are looking for you are in the wrong place – but the effort required is well worth the results you get. Remember this isn't about being perfect, no one is, this is about building new habits and taking the right steps.

You have already taken your first step to working on your health and thank you for making that step reading this book. I hope that now you will go out there and implement the strategies into your life if you haven't started this already. Let's recap.

Take The MitoImmune Health Assessment online and find out which pillars of your health need support and which you are doing well with. Start with The MitoImmune Nutrition Plan – this impacts all of the pillars supporting your health. Find your lowest scoring areas on the assessment and work to add in the specific nutrients to support these pillars. Work through the 4 Lifestyle S's, implementing steps to support your sleep, stress, strength and social connection. If you feel overwhelmed remember small steps equal big change, take it one step at a time. Don't forget you have all the tools in the online book resources to help support you through these steps. There are also meal plans, shopping lists and recipes for you to access.

If either you want to jump in and dig into root causes right now, or after implementing these steps in this book you feel the benefits, but you want to further your journey, it's time to reach out (see Appendix I). I can help to advise you on the most beneficial route for you personally.

You might be wondering what working with a functional medicine practitioner is like. What you get in clinic that you don't get in a book is a full case review prior to your initial appointment. I go through some forms you fill in and your previous medical history as well as previous test results. We then have a full functional health assessment, spending 90 minutes together on your health – consider how much we can go through in this time when you normally get around 10 minutes with your doctor. I then carry out science led research into your case, we action any functional testing and we start to build your protocol. I use my clinical training and experience to provide you with personalized care. I review and interpret your functional test results and we adapt your plans to further personalize your healthcare.

What is important to remember here is that it doesn't have to be either/or. Now you know what functional medicine is and how it can impact your health you may want to reap the benefits and give it a try, but you don't have to choose one or the other. You don't have to take a conventional medicine or a functional medicine route, in fact in many cases I work in collaboration with conventional medicine doctors, either through doctors you are already under, your GP or we can help you get a referral.

Remember this is about extending your healthspan. We know that our food and lifestyle choices impact our health both now and in the future. Leaving chronic stressors unchecked, whether physical, or emotional, leads to burnout, which leads to chronic health conditions. Protecting our energy and building immune resilience is key to building up your body to not only protect your future, but to let you be the best version of YOU.

Health is a journey, working on your health in a holistic way to benefit you both now and in the future is not a quick fix, it is a lifestyle, one that will benefit you for the rest of your life, which will not only be longer but also healthier, more active, filled with success and the ability to do all the things you love.

Good luck on your health journey and I hope you find the path that works for you.

Nicole Goode

2024

APPENDIX I

Resources

Nicole's websites:

- nicolegoodehealth.com
- goodehealth.uk
- optimalyoubook.com

Access Online Book Resources: www.nicolegoodehealth.com/optimal-you-book-resources. Password: OptimalYouBook. In the resources you can access all the tools we talk about in the book, recipes to help you move through the MitoImmune Plan, supplement recommendations and more.

To take the Free MitoImmune Health Assessment: www.nicolegoodehealth.com/mitoimmune-health-assessment

If you would like to book a clinic appointment , or free enquiry call, visit goodehealth.uk

Virtual appointments can support you wherever you are in the world.

APPENDIX II

My favourites

Books

- *Blue Zones: Secrets for Living Longer* by Dan Buettner
  I'm fascinated by the science behind the blue zones. Dan's book brilliantly discusses all the lessons we can learn from the blue zones and it is also a beautiful book with wonderful photography, one to leave out on the coffee table.
- *Breath: The New Science of a Lost Art* by James Nestor
  A book that will truly change how you think about the connection between our breath, our body and our mind.
- *Breathe In Breathe Out* by Stuart Sandeman
  Stuart Sandeman is a breathwork expert and if you haven't joined his live Instagram sessions I highly recommend you do. Both Stuart's book and the sessions he runs have been a huge part of my mindfulness journey.
- *Four Pillar Plan* by Dr Rangan Chatterjee
  Remember when we said small changes can lead to big differences? Well Dr Chatterjee is brilliant at this. His first book has some brilliant actionable steps. I particularly like his simple 'workouts' and exercise advice which I regularly advise patients to go and read.
- *Miracle Morning* by Hal Elrod
  If you want to change how you view your mornings this is the book for you, even as someone like myself who is a definite night owl and not an early bird, this book can change how you view your morning routine to bring productiveness with a sense of calm.
- *Nurture* by Carole Bamford
  Carole Bamford is the Queen of organic farming, way before it was popular. Before talk of pesticides and our health was commonplace, Carole was already fighting for organic and sustainable farming. Her knowledge

and passion shines through in this beautiful book – another coffee table favourite.

- *The Stress Solution* by Dr Rangan Chatterjee
  Another of Dr Chatterjee's books, we've looked at the importance of stress management and if you are wanting to take steps to start your journey to a life where stress is better managed, this is the place to start.

Cookbooks

- *Get the Glow* by Madeleine Shaw
  This is Madeleine's first book and it's still a staple in my kitchen. All her books are great and I use them all regularly, but if you want to get started with one I would still reach for this every time. It's full of tasty, simple and nutritious recipes.
- *Healthy Made Simple* by Ella Mills (Deliciously Ella)
  Again Ella has lots of cookbooks and they are all good, but this new one is packed with recipes that can be ready in 30 minutes or less, which I think is a real win after a full day of work. Its recipes like this that really help us to stay on track with our health.
- *One Pot Pan Planet* by Anna Jones
  If you love vegetarian food or you want to get more plant-based meals into your week this is THE book for you. Everything is a one pot or pan recipe which makes cooking a breeze and washing up even easier, but more importantly the food is delicious. Many of her recipes have become weekly staples in my house.
- *Plenty* by Yotam Ottolenghi
  If you are looking to get that diversity into your diet and eat more plant-based meals, this is how you take it to the next level. While some of the recipes are a little more complex, it's well worth the work. His plant-based food is tasty enough that even meat lovers won't be able to refuse it.
- *Riverfood Cookbooks* www.riverford.co.uk/essentials/books-and-gifts
  This is a seasonal release and I love how they focus on quality organic foods that are in season to support your body at the right time of year. As I write this the Spring Summer version is available; as we go to print the Autumn Winter version will be out. Get both!
- *Tapas Revolution and Spanish Made Simple* by Omar Allibhoy
  I partially grew up on the beautiful island of Mallorca, it's my second home and therefore I love Spanish food, although I find it can be hard to come by good Spanish cuisine in the UK, which is a shame as it's full of the delights of a Mediterranean diet. Omar Allibhoy is changing that. His books contain great food, full of that Spanish flavour, so that anyone can

make it at home. *Tapas Revolution*, in particular, contains many staple recipes in our house.

- *The Doctor's Kitchen: Eat to Beat Illness* by Dr Rupy Aujla
  With great insights into eating to live healthy, there are also plenty of great recipes. I find Dr Rupy's recipes to be simple and easy to make without sacrificing taste.
- *The Tucci Cookbook* by Stanley Tucci
  Who doesn't love Italian food (and who doesn't love Stanley Tucci)? Tucci's Italian adventure was brilliant TV centred around food but also importantly, as we have discussed in this book, the social connection that can come with good food. Stanley's book is full of family friendly recipes that will make you want to gather all your friends up to enjoy them with you.

Films/Programmes

- *Live to 100: Secrets of the Blue Zones*, Netflix
  A fascinating insight into the blue zones and how the way of life can impact longevity, but for me the really interesting part is in seeing how they create a blue zone; this really solidifies the importance of nutrition and lifestyle for our health.
- *Common Ground Film*, Netflix; and *Kiss the Ground Film*, Netflix
  Both produced by the same team. If you want to learn about the quality of our food, sustainable farming and importantly soil health, these are the films to watch. I promise you they will change the way you think about your food forever.
- *Seaspiracy*, Netflix
  A hard watch at times, but one we should learn about to understand what is happening in our oceans and to our food.

Tools

- Air doctor HEPA air purifier: for discount use the link: https://amazing-air.co.uk/products/amazing-air-3000?oid=5&affid=343, or visit the online book resources.
- Sensate neuromodulation for anxiety and stress relief: https://getsensate.com/NGHealth CODE: NGHealth for discount.
- The Othership YouTube channel and app for meditations, Nervous System Reset is a favourite of mine: www.youtube.com/watch?v=qlTC2HBmPeM
- The dirty dozen and clean fifteen list: https://www.ewg.org/foodnews/full-list.php
- Access the online book resources for more tools.

Supplements

For discounts on my favourite supplement brands visit the online book resources at www.nicolegoodehealth.com/optimal-you-book-resources; Password: OptimalYouBook

'No Nasties' skincare/cosmetic brands

- Bamford
  Sustainable, organic skincare inspired by nature, I'm a fan of all of the products. www.bamford.com/beauty.html
- Medik8
  British Dermatology tested science backed skincare, particularly good for sorting out dry skin and retinal products (instead of retinol). www.medik8.com/
- Hanna Sillitoe
  Created for people with skin conditions like eczema and psoriasis. www.hannasillitoe.com/
- Westman Atelier
  Makeup created with scientists that benefits the skin, I especially like the foundation, contour and blushes. www.westman-atelier.com/en-gb
- Victoria Beckham Beauty
  Clean and sustainable makeup, I'm a fan of the eye makeup. https://victoriabeckhambeauty.com/en-gb
- Hourglass Cosmetics
  Vegan makeup and skincare, cruelty free, great for the lips. www.hourglasscosmetics.co.uk/
- Salt and Stone
  The best of the natural deodorants, this actually works. www.saltandstone.com/en-gb

APPENDIX III

References

Part 1: The foundations of optimal health

- Ward BW, Schiller JS, Goodman RA. Multiple chronic conditions among US adults: a 2012 update. Prev Chronic Dis. 2014 Apr 17;11:E62. doi: 10.5888/pcd11.130389. PMID: 24742395; PMCID: PMC3992293. https://pubmed.ncbi.nlm.nih.gov/24742395/
- Wang et al. Global, regional, and national life expectancy, all-cause mortality, and cause-specific mortality for 249 causes of death, 1980–2015: a systematic analysis for the Global Burden of Disease Study 2015. The Lancet. 2016;388(10053):1459–544. doi: 10.1016/S0140-6736(16)31012-1. www.thelancet.com/journals/lancet/article/PIIS0140-6736(16)31012-1/fulltext
- Leroy L, Bayliss E, Domino M, Miller BF, Rust G, Gerteis J, Miller T; AHRQ MCC Research Network. The Agency for Healthcare Research and Quality Multiple Chronic Conditions Research Network: overview of research contributions and future priorities. Med Care. 2014 Mar;52 Suppl 3:S15-22. doi: 10.1097/MLR.0000000000000095. PMID: 24561753. https://pubmed.ncbi.nlm.nih.gov/24561753/
- Centers for Medicare and Medicaid services. NHE fact sheet. www.cms.gov/data-research/statistics-trends-and-reports/national-health-expenditure-data/nhe-fact-sheet
- Whoop. 'The Cost of Business Burnout'. www.whoopunite.com/blog/business/articles/cost-of-employee-burnout/
- AXA. 'The True Cost of Running on Empty: work-related stress costing UK economy £28bn a year'. www.axa.co.uk/newsroom/media-releases/2023/the-true-cost-of-running-on-empty-work-related-stress-costing-uk-economy-28bn-a-year/
- WHO. Global spending on health: rising to the pandemic's challenges. www.who.int/publications/i/item/9789240064911#

- IFM. Why Functional Medicine Matters: The Problem of Health Care: High Cost and Dependency. www.ifm.org/functional-medicine/why-functional-medicine-matters/
- CDC. Chronic Diseases in America. www.cdc.gov/chronicdisease/resources/infographic/chronic-diseases.htm
- Nattagh-Eshtivani E, Sani MA, Dahri M, Ghalichi F, Ghavami A, Arjang P, Tarighat-Esfanjani A. The role of nutrients in the pathogenesis and treatment of migraine headaches: Review. Biomed Pharmacother. 2018 Jun; 102:317–325. doi: 10.1016/j.biopha.2018.03.059. Epub 2018 Mar 22. PMID: 29571016. https://pubmed.ncbi.nlm.nih.gov/29571016/
- Steelman A. Infection as an environmental trigger of multiple sclerosis disease exacerbation. Frontiers in Immunology, 19 October 2015 Sec. Multiple Sclerosis and Neuroimmunology, Volume 6, 2015 doi: 10.3389/fimmu.2015.00520. www.frontiersin.org/journals/immunology/articles/10.3389/fimmu.2015.00520/full
- Linden J et al. Clostridium perfringens epsilon toxin causes selective death of mature oligodendrocytes and central nervous system demyelination. ASM Journals, mBio, Vol. 6, No. 3, 16 June 2015. doi: https://doi.org/10.1128/mbio.02513-14 https://journals.asm.org/doi/full/10.1128/mbio.02513-14
- Johnson KB, Wei WQ, Weeraratne D, Frisse ME, Misulis K, Rhee K, Zhao J, Snowdon JL. Precision medicine, AI, and the future of personalized health care. Clin Transl Sci. 2021 Jan;14(1):86–93. doi: 10.1111/cts.12884. Epub 2020 Oct 12. PMID: 32961010; PMCID: PMC7877825. www.ncbi.nlm.nih.gov/pmc/articles/PMC7877825/
- National Institute of Aging. 'Alzheimer's disease genetics fact sheet'. www.nia.nih.gov/health/genetics-and-family-history/alzheimers-disease-genetics-fact-sheet#:~:text=
- Alzheimer's Association. 'Is Alzheimer's genetic?'. www.alz.org/alzheimers-dementia/what-is-alzheimers/causes-and-risk-factors/genetics
- Furman D, Campisi J, Verdin E, Carrera-Bastos P, Targ S, Franceschi C, Ferrucci L, Gilroy DW, Fasano A, Miller GW, Miller AH, Mantovani A, Weyand CM, Barzilai N, Goronzy JJ, Rando TA, Effros RB, Lucia A, Kleinstreuer N, Slavich GM. Chronic inflammation in the etiology of disease across the life span. Nat Med. 2019 Dec;25(12):1822–1832. doi: 10.1038/s41591-019-0675-0. Epub 2019 Dec 5. PMID: 31806905; PMCID: PMC7147972. https://pubmed.ncbi.nlm.nih.gov/31806905/
- GBD 2017 Diet Collaborators. Health effects of dietary risks in 195 countries, 1990–2017: a systematic analysis for the Global Burden of Disease Study 2017. The Lancet. 2019; 393: 1958–72. April 3, 2019. doi: 10.1016/S0140-6736(19)30041-8 www.thelancet.com/journals/lancet/article/PIIS0140-6736(19)30041-8/fulltext#seccestitle160

Part 2: The pillars of optimal health

Mitochondria

- Brand MD, Orr AL, Perevoshchikova IV, Quinlan CL. The role of mitochondrial function and cellular bioenergetics in ageing and disease. Br J Dermatol. 2013 July;169 Suppl 2(02):1–8. doi: 10.1111/bjd.12208. PMID: 23786614; PMCID: PMC4321783. www.ncbi.nlm.nih.gov/pmc/articles/PMC4321783/
- Pizzorno J. Mitochondria-Fundamental to Life and Health. Integr Med (Encinitas). 2014 Apr;13(2):8–15. PMID: 26770084; PMCID: PMC4684129. www.ncbi.nlm.nih.gov/pmc/articles/PMC4684129/
- Apostolova N, Victor VM. Molecular strategies for targeting antioxidants to mitochondria: therapeutic implications. Antioxid Redox Signal. 2015 Mar 10;22(8):686–729. doi: 10.1089/ars.2014.5952. PMID: 25546574; PMCID: PMC4350006. www.ncbi.nlm.nih.gov/pmc/articles/PMC4350006/
- Tahrir FG, Langford D, Amini S, Mohseni Ahooyi T, Khalili K. Mitochondrial quality control in cardiac cells: Mechanisms and role in cardiac cell injury and disease. J Cell Physiol. 2019 June;234(6):8122–8133. doi: 10.1002/jcp.27597. Epub 2018 Nov 11. PMID: 30417391; PMCID: PMC6395499. www.ncbi.nlm.nih.gov/pmc/articles/PMC6395499/
- Anqi Li et al., Mitochondrial dynamics in adult cardiomyocytes and heart diseases. Front. Cell Dev. Biol., 17 December 2020. Sec. Mitochondrial Research, Volume 8 – 2020 | https://doi.org/10.3389/fcell.2020.584800

Immune

- Nathalie Conrad, PhD et al. Incidence, prevalence, and co-occurrence of autoimmune disorders over time and by age, sex, and socioeconomic status: a population-based cohort study of 22 million individuals in the UK. The Lancet. Volume 401, Issue 10391, 1878–1890, 3 June 2023. www.thelancet.com/journals/lancet/article/PIIS0140-6736(23)00457-9/
- Gallagher, L., Autoimmune disorders affect about one in ten individuals. Imperial. 9 May 2023. www.imperial.ac.uk/news/244793/autoimmune-disorders-affect-about-individuals
- Conrad N, Misra S, Verbakel JY, Verbeke G, Molenberghs G, Taylor PN, Mason J, Sattar N, McMurray JJV, McInnes IB, Khunti K, Cambridge G. Incidence, prevalence, and co-occurrence of autoimmune disorders over time and by age, sex, and socioeconomic status: a population-based cohort study of 22 million individuals in the UK. Lancet. 2023 June 3;401(10391):1878–1890. doi: 10.1016/S0140-6736(23)00457-9.

Epub 2023 May 5. PMID: 37156255. https://pubmed.ncbi.nlm.nih.gov/37156255/

- The Guardian. Global spread of autoimmune disease blamed on western diet. 2022. www.theguardian.com/science/2022/jan/08/global-spread-of-autoimmune-disease-blamed-on-western-diet
- Wiersinga WM. (2018). Hashimoto's thyroiditis. In: Vitti, P., Hegedüs, L. (eds) Thyroid Diseases. Endocrinology. Springer, Cham. https://doi.org/10.1007/978-3-319-45013-1_7. https://link.springer.com/referenceworkentry/10.1007/978-3-319-45013-1_7
- National Institute on Aging. Immune resilience is key to a long and healthy life. 30 June 2023. www.nia.nih.gov/news/immune-resilience-key-long-and-healthy-life
- Autoimmune Association. AutoImmune Disease List. https://autoimmune.org/disease-information/
- Kennedy MA. A brief review of the basics of immunology: the innate and adaptive response. Vet Clin North Am Small Anim Pract. 2010 May;40(3):369–379. doi: 10.1016/j.cvsm.2010.01.003. PMID: 20471522. https://pubmed.ncbi.nlm.nih.gov/20471522/

Brain

- Sabayan B et al. The role of population-level preventive care for brain health in ageing. The Lancet Healthy Longevity. Volume 4, Issue 6, e274–e283, June 2023. DOI:https://doi.org/10.1016/S2666-7568(23)00051-X www.thelancet.com/journals/lanhl/article/PIIS2666-7568(23)00051-X/fulltext
- Wang Y, Pan Y, Li H. What is brain health and why is it important? BMJ 2020; 371 :m3683 doi:10.1136/bmj.m3683. www.bmj.com/content/371/bmj.m3683
- Peters R. Ageing and the brain. Postgrad Med J. 2006 Feb;82(964):84–88. doi: 10.1136/pgmj.2005.036665. PMID: 16461469; PMCID: PMC2596698. https://www.ncbi.nlm.nih.gov/pmc/articles/PMC2596698/
- Svard L. Learn an instrument—change your brain. The Musical Brain: What Students, Teachers, and Performers Need to Know (New York, 2023; online edn, Oxford Academic, 23 Feb. 2023), https://doi.org/10.1093/oso/9780197584170.003.0004, accessed 10 May 2024. https://academic.oup.com/book/45551/chapter-abstract/394680087?redirectedFrom=fulltext
- Stern Y, Barnes CA, Grady C, Jones RN, Raz N. Brain reserve, cognitive reserve, compensation, and maintenance: operationalization, validity, and mechanisms of cognitive resilience. Neurobiol Aging. 2019 Nov;83:124–

129. doi: 10.1016/j.neurobiolaging.2019.03.022. PMID: 31732015; PMCID: PMC6859943. www.ncbi.nlm.nih.gov/pmc/articles/PMC6859943/

- Gulati A. Understanding neurogenesis in the adult human brain. Indian J Pharmacol. 2015 Nov-Dec;47(6):583–584. doi: 10.4103/0253-7613.169598. PMID: 26729946; PMCID: PMC4689008. www.ncbi.nlm.nih.gov/pmc/articles/PMC4689008/
- Herculano-Houzel S. Front. Hum. Neurosci., 09 November 2009 Sec. Cognitive Neuroscience Volume 3 – 2009 | https://doi.org/10.3389/neuro.09.031.2009 www.frontiersin.org/articles/10.3389/neuro.09.031.2009/full
- Chang CY, Ke DS, Chen JY. Essential fatty acids and human brain. Acta Neurol Taiwan. 2009 Dec;18(4):231–241. PMID: 20329590. https://pubmed.ncbi.nlm.nih.gov/20329590/
- Kaur N, Chugh H, Sakharkar MK, Dhawan U, Chidambaram SB, Chandra R. Neuroinflammation mechanisms and phytotherapeutic intervention: a systematic review. ACS Chem Neurosci. 2020 Nov 18;11(22):3707–3731. doi: 10.1021/acschemneuro.0c00427. Epub 2020 Nov 4. PMID: 33146995. https://pubmed.ncbi.nlm.nih.gov/33146995/

Adrenal

- Rosol TJ, Yarrington JT, Latendresse J, Capen CC. Adrenal gland: structure, function, and mechanisms of toxicity. Toxicol Pathol. 2001 Jan-Feb;29(1):41–48. doi: 10.1080/019262301301418847. PMID: 11215683. https://pubmed.ncbi.nlm.nih.gov/11215683/
- Tohei A. Studies on the functional relationship between thyroid, adrenal and gonadal hormones. J Reprod Dev. 2004 Feb;50(1):9–20. doi: 10.1262/jrd.50.9. PMID: 15007197. https://pubmed.ncbi.nlm.nih.gov/15007197/
- Armario A. The hypothalamic-pituitary-adrenal axis: what can it tell us about stressors? CNS Neurol Disord Drug Targets. 2006 Oct;5(5):485–501. doi: 10.2174/187152706778559336. PMID: 17073652. https://pubmed.ncbi.nlm.nih.gov/17073652/
- Stojanovich L, Marisavljevich D. Stress as a trigger of autoimmune disease. Autoimmun Rev. 2008 Jan;7(3):209–213. doi: 10.1016/j.autrev.2007.11.007. Epub 2007 Nov 29. PMID: 18190880. https://pubmed.ncbi.nlm.nih.gov/18190880/
- Juruena MF, Eror F, Cleare AJ, Young AH. The role of early life stress in HPA axis and anxiety. Adv Exp Med Biol. 2020;1191:141–153. doi: 10.1007/978-981-32-9705-0_9. PMID: 32002927. https://pubmed.ncbi.nlm.nih.gov/32002927/
- Joseph DN, Whirledge S. Stress and the HPA axis: balancing homeostasis and fertility. Int J Mol Sci. 2017 Oct 24;18(10):2224. doi: 10.3390/ijms18102224. PMID: 29064426; PMCID: PMC5666903. https://pubmed.ncbi.nlm.nih.gov/29064426/

- Misiak B, Łoniewski I, Marlicz W, Frydecka D, Szulc A, Rudzki L, Samochowiec J. The HPA axis dysregulation in severe mental illness: can we shift the blame to gut microbiota? Prog Neuropsychopharmacol Biol Psychiatry. 2020 Aug 30;102:109951. doi: 10.1016/j.pnpbp.2020.109951. Epub 2020 Apr 23. PMID: 32335265. https://pubmed.ncbi.nlm.nih.gov/32335265/
- Cheiran Pereira G, Piton E, Moreira Dos Santos B, Ramanzini LG, Muniz Camargo LF, Menezes da Silva R, Bochi GV. Microglia and HPA axis in depression: an overview of participation and relationship. World J Biol Psychiatry. 2022 Mar;23(3):165–182. doi: 10.1080/15622975.2021.1939154. Epub 2021 Jul 7. PMID: 34100334. https://pubmed.ncbi.nlm.nih.gov/34100334/

Gut

- National Institutes of Health. NIH Human Microbiome Project defines normal bacterial makeup of the body. Wednesday, 13 June 2012. www.nih.gov/news-events/news-releases/nih-human-microbiome-project-defines-normal-bacterial-makeup-body
- Wu HJ, Wu E. The role of gut microbiota in immune homeostasis and autoimmunity. Gut Microbes. 2012 Jan-Feb;3(1):4–14. doi: 10.4161/gmic.19320. Epub 2012 Jan 1. PMID: 22356853; PMCID: PMC3337124. www.ncbi.nlm.nih.gov/pmc/articles/PMC3337124/
- Wiertsema SP, van Bergenhenegouwen J, Garssen J, Knippels LMJ. The interplay between the gut microbiome and the immune system in the context of infectious diseases throughout life and the role of nutrition in optimizing treatment strategies. Nutrients. 2021 Mar 9;13(3):886. doi: 10.3390/nu13030886. PMID: 33803407; PMCID: PMC8001875. www.ncbi.nlm.nih.gov/pmc/articles/PMC8001875/
- Tomasello G, Mazzola M, Jurjus A, Cappello F, Carini F, Damiani P, Gerges Geagea A, Zeenny MN, Leone A. The fingerprint of the human gastrointestinal tract microbiota: a hypothesis of molecular mapping. J Biol Regul Homeost Agents. 2017 Jan–Mar;31(1):245–249. PMID: 28337900. https://pubmed.ncbi.nlm.nih.gov/28337900/
- Breit S, Kupferberg A, Rogler G, Hasler G. Vagus nerve as modulator of the brain-gut axis in psychiatric and inflammatory disorders. Front Psychiatry. 2018 Mar 13;9:44. doi: 10.3389/fpsyt.2018.00044. PMID: 29593576; PMCID: PMC5859128. www.ncbi.nlm.nih.gov/pmc/articles/PMC5859128/
- Cavaliere G, Catapano A, Trinchese G, Cimmino F, Penna E, Pizzella A, Cristiano C, Lama A, Crispino M, Mollica MP. Butyrate improves neuroinflammation and mitochondrial impairment in cerebral cortex and

synaptic fraction in an animal model of diet-induced obesity. Antioxidants 2023, 12, 4. https://doi.org/10.3390/antiox12010004

- Conder E et al. Role of the gut microbiome and pathogens in immune and inflammatory diseases. April 2023. www.gastrojournal.org/article/S0016-5085(23)00425-0/pdf
- Spencer NJ, Hu H. Enteric nervous system: sensory transduction, neural circuits and gastrointestinal motility. Nat Rev Gastroenterol Hepatol. 2020 Jun;17(6):338–351. doi: 10.1038/s41575-020-0271-2. Epub 2020 Mar 9. PMID: 32152479; PMCID: PMC7474470. www.ncbi.nlm.nih.gov/pmc/articles/PMC7474470/
- Pargin, E. et al., The human gut virome: composition, colonization, interactions, and impacts on human health. Front. Microbiol., 24 May 2023 Sec. Virology, Volume 14 2023 | https://doi.org/10.3389/fmicb.2023.963173 www.frontiersin.org/journals/microbiology/articles/10.3389/fmicb.2023.963173/full
- Fasano A. Leaky gut and autoimmune diseases. Clin Rev Allergy Immunol. 2012 Feb;42(1):71–78. doi: 10.1007/s12016-011-8291-x. PMID: 2210 9896. https://pubmed.ncbi.nlm.nih.gov/22109896/
- Ilchmann-Diounou H, Menard S. Psychological stress, intestinal barrier dysfunctions, and autoimmune disorders: an overview. Front Immunol. 2020 Aug 25;11:1823. doi: 10.3389/fimmu.2020.01823. PMID: 3298 3091; PMCID: PMC7477358. https://pubmed.ncbi.nlm.nih.gov/32983091/

Thyroid

- NICE. May 2021. Hypothyroidism: How common is it?. https://cks.nice.org.uk/topics/hypothyroidism/background-information/prevalence/
- Office on Women's Health. Thyroid disease. Feb. 2021. www.womenshealth.gov/a-z-topics/thyroid-disease
- Gesing A. The thyroid gland and the process of aging. Thyroid Res. 2015 Jun 22;8(Suppl 1):A8. doi: 10.1186/1756-6614-8-S1-A8. PMCID: PMC4480281. www.ncbi.nlm.nih.gov/pmc/articles/PMC4480281/
- Farebrother J, Zimmermann MB, Andersson M. Excess iodine intake: sources, assessment, and effects on thyroid function. Ann N Y Acad Sci. 2019 Jun;1446(1):44–65. doi: 10.1111/nyas.14041. Epub 2019 Mar 20. PMID: 30891786. https://pubmed.ncbi.nlm.nih.gov/30891786/
- National Institutes of Health. Iodine. May 2024. https://ods.od.nih.gov/factsheets/Iodine-HealthProfessional/
- Zaletel K, Gaberšček S. Hashimoto's thyroiditis: from genes to the disease. Curr Genomics. 2011 Dec;12(8):576–588. doi: 10.2174/138920211798120763. PMID: 22654557; PMCID: PMC3271310. https://pubmed.ncbi.nlm.nih.gov/22654557/

- Hashemipour M et al. High prevalence of goiter in an iodine replete area. Asia Pac J Clin Nutr, 2007: 16(3), 403–410. www.researchgate.net/publication/6135806_High_prevalence_of_goiter_in_an_iodine_replete_area_Do_thyroid_auto-antibodies_play_a_role
- Liu J et al. Excessive iodine promotes pyroptosis of thyroid follicular epithelial cells in Hashimoto's thyroiditis through the ROS-NF-κB-NLRP3 pathway. Front. Endocrinol., 20 November 2019 Sec. Thyroid Endocrinology Volume 10 - 2019. https://doi.org/10.3389/fendo.2019.00778
- Teng W, Shan Z, Teng X, Guan H, Li Y, Teng D, Jin Y, Yu X, Fan C, Chong W, Yang F, Dai H, Yu Y, Li J, Chen Y, Zhao D, Shi X, Hu F, Mao J, Gu X, Yang R, Tong Y, Wang W, Gao T, Li C. Effect of iodine intake on thyroid diseases in China. N Engl J Med. 2006 Jun 29;354(26):2783–2793. doi: 10.1056/NEJMoa054022. PMID: 16807415. https://pubmed.ncbi.nlm.nih.gov/16807415/
- Reinhardt W et al. Effect of small doses of iodine on thyroid function in patients with Hashimoto's thyroiditis residing in an area of mild iodine deficiency. European Journal of Endocrinology, 1998: 139, 23-28. doi:10.1530/eje.0.1390023 www.researchgate.net/publication/13582669_Effect_of_small_doses_of_iodine_on_thyroid_function_in_patients_with_Hashimoto's_thyroiditis_residing_in_an_area_of_mild_iodine_deficiency
- Ferrari SM, Fallahi P, Antonelli A, Benvenga S. Environmental issues in thyroid diseases. Front Endocrinol (Lausanne). 2017 Mar 20;8:50. doi: 10.3389/fendo.2017.00050. PMID: 28373861; PMCID: PMC5357628. www.ncbi.nlm.nih.gov/pmc/articles/PMC5357628/
- Leemans M, Couderq S, Demeneix B, Fini JB. Pesticides with potential thyroid hormone-disrupting effects: a review of recent data. Front Endocrinol (Lausanne). 2019 Dec 9;10:743. doi: 10.3389/fendo.2019.00743. PMID: 31920955; PMCID: PMC6915086. www.ncbi.nlm.nih.gov/pmc/articles/PMC6915086/
- Fröhlich E, Wahl R. Thyroid autoimmunity: role of anti-thyroid antibodies in thyroid and extra-thyroidal diseases. Front. Immunol. 8:521. 2017. doi: 10.3389/fimmu.2017.00521
- Mullur R, Liu Y-Y, Brent GA. Thyroid hormone regulation of metabolism. Physiol Rev 94: 355–382, 2014. doi:10.1152/physrev.00030.2013
- Vojdani A. Molecular mimicry as a mechanism for food immune reactivities and autoimmunity. Altern Ther Health Med. 2015;21 Suppl 1:34–45. PMID: 25599184. https://pubmed.ncbi.nlm.nih.gov/25599184/

Hormone

- Hodges RE, Minich DM. Modulation of metabolic detoxification pathways using foods and food-derived components: a scientific review with clinical application. J Nutr Metab. 2015;2015:760689. doi: 10.1155/2015/760689. Epub 2015 Jun 16. PMID: 26167297; PMCID: PMC4488002. www.ncbi.nlm.nih.gov/pmc/articles/PMC4488002/

- Ross GH, Sternquist MC. Methamphetamine exposure and chronic illness in police officers: significant improvement with sauna-based detoxification therapy. Toxicology and Industrial Health. 2012;28(8):758–768. doi:10.1177/0748233711425070
- IFM. Supporting liver function with nutrition. www.ifm.org/news-insights/detox-food-plan/
- Hodges RE, Minich DM. Modulation of metabolic detoxification pathways using foods and food-derived components: a scientific review with clinical application. J Nutr Metab. 2015;2015:760689. doi:1155/2015/760689
- Jackson SJ, Singletary KW, Murphy LL, Venema RC, Young AJ. Phytonutrients differentially stimulate NAD(P)H:quinone oxidoreductase, inhibit proliferation, and trigger mitotic catastrophe in hepa1c1c7 cells. J Med Food. 2016;19(1):47–53. doi:1089/jmf.2015.0079
- Abbaoui B, Lucas CR, Riedl KM, Clinton SK, Mortazavi A. Cruciferous vegetables, isothiocyanates and bladder cancer prevention. Mol Nutr Food Res. 2018;62(18):e1800079. doi:1002/mnfr.201800079
- Jiang X, Liu Y, Ma L, et al. Chemopreventive activity of sulforaphane. Drug Des Devel Ther. 2018;12:2905–2913. doi:2147/DDDT.S100534
- Menezo Y, Clement P, Clement A, Elder K. Methylation: an ineluctable biochemical and physiological process essential to the transmission of life. Int J Mol Sci. 2020 Dec 7;21(23):9311. doi: 10.3390/ijms21239311. PMID: 33297303; PMCID: PMC7730869. www.ncbi.nlm.nih.gov/pmc/articles/PMC7730869/
- Lanata CM, Chung SA, Criswell LA. DNA methylation 101: what is important to know about DNA methylation and its role in SLE risk and disease heterogeneity. Lupus Sci Med. 2018 Jul 25;5(1):e000285. doi: 10.1136/lupus-2018-000285. PMID: 30094041; PMCID: PMC6069928. www.ncbi.nlm.nih.gov/pmc/articles/PMC6069928/
- Łoboś P, Regulska-Ilow B. Link between methyl nutrients and the DNA methylation process in the course of selected diseases in adults. Rocz Panstw Zakl Hig. 2021;72(2):123–136. doi: 10.32394/rpzh.2021.0157. PMID: 34114759. https://pubmed.ncbi.nlm.nih.gov/34114759/

Cardiometabolic

- Xu C, Zhang P, Cao Z. Cardiovascular health and healthy longevity in people with and without cardiometabolic disease: a prospective cohort study. EClinicalMedicine. 2022 Mar 6;45:101329. doi: 10.1016/j.eclinm.2022.101329. PMID: 35284807; PMCID: PMC8904213. www.ncbi.nlm.nih.gov/pmc/articles/PMC8904213/

- Lagström H et al. Diet quality as a predictor of cardiometabolic disease–free life expectancy: the Whitehall II cohort study, The American Journal of Clinical Nutrition, Volume 111, Issue 4, 2020, 787–794. https://doi.org/10.1093/ajcn/nqz329. www.sciencedirect.com/science/article/pii/S0002916522010656
- Jin Y et al. Cardiometabolic multimorbidity, lifestyle behaviours, and cognitive function: a multicohort study. Volume 4, Issue 6, e265–e273, June 2023. DOI:https://doi.org/10.1016/S2666-7568(23)00054-5 www.thelancet.com/journals/lanhl/article/PIIS2666-7568(23)00054-5/fulltext
- National Academies of Sciences, Engineering, and Medicine; Division of Behavioral and Social Sciences and Education; Committee on National Statistics; Committee on Population; Committee on Rising Midlife Mortality Rates and Socioeconomic Disparities; Becker T, Majmundar MK, Harris KM, editors. High and rising mortality rates among working-age adults. Washington (DC): National Academies Press (US); 2021 Mar 2. 9, Cardiometabolic Diseases. Available from: www.ncbi.nlm.nih.gov/books/NBK571925/
- WHO. Cardiovascular diseases (CVDs). 11 June 2021. www.who.int/news-room/fact-sheets/detail/cardiovascular-diseases-(cvds)
- Wang Z, Chen J, Zhu L, Jiao S, Chen Y, Sun Y. Metabolic disorders and risk of cardiovascular diseases: a two-sample mendelian randomization study. BMC Cardiovasc Disord. 2023 Oct 31;23(1):529. doi: 10.1186/s12872-023-03567-3. PMID: 37907844; PMCID: PMC10617200. www.ncbi.nlm.nih.gov/pmc/articles/PMC10617200/
- American Diabetes Association. Statistics about diabetes. https://diabetes.org/about-diabetes/statistics/about-diabetes
- Diabetes UK. Number of people living with diabetes in the uk tops 5 million for the first time. April 2023. www.diabetes.org.uk/about-us/news-and-views/number-people-living-diabetes-uk-tops-5-million-first-time
- Guess N. The growing use of continuous glucose monitors in people without diabetes: an evidence-free zone. Practical Diabetes. Volume40, Issue5, September/October 2023, 19–22a02 https://wchh.onlinelibrary.wiley.com/doi/full/10.1002/pdi.2475

Part 3: The MitoImmune Way – your path to optimal health

Nutrition and nutrients

- Du J, Zhu M, Bao H, Li B, Dong Y, Xiao C, Zhang GY, Henter I, Rudorfer M, Vitiello B. The role of nutrients in protecting mitochondrial function and neurotransmitter signaling: implications for the treatment of depression,

PTSD, and suicidal behaviors. Crit Rev Food Sci Nutr. 2016 Nov 17;56(15):2560–2578. doi: 10.1080/10408398.2013.876960. PMID: 25365455; PMCID: PMC4417658. www.ncbi.nlm.nih.gov/pmc/articles/PMC4417658/

- Pizzorno J. Mitochondria-fundamental to life and health. Integr Med (Encinitas). 2014 Apr;13(2):8–15. PMID: 26770084; PMCID: PMC4684129. www.ncbi.nlm.nih.gov/pmc/articles/PMC4684129/
- Calder PC. Foods to deliver immune-supporting nutrients. Curr Opin Food Sci. 2022 Feb;43:136–145. doi: 10.1016/j.cofs.2021.12.006. Epub 2021 Dec 18. PMID: 34976746; PMCID: PMC8702655. www.ncbi.nlm.nih.gov/pmc/articles/PMC8702655/
- Childs CE, Calder PC, Miles EA. Diet and immune function. Nutrients. 2019 Aug 16;11(8):1933. doi: 10.3390/nu11081933. PMID: 31426423; PMCID: PMC6723551. www.ncbi.nlm.nih.gov/pmc/articles/PMC6723551/
- Gómez-Pinilla F. Brain foods: the effects of nutrients on brain function. Nat Rev Neurosci. 2008 Jul;9(7):568–578. doi: 10.1038/nrn2421. PMID: 18568016; PMCID: PMC2805706. www.ncbi.nlm.nih.gov/pmc/articles/PMC2805706/
- Patani A, Balram D, Yadav VK, Lian KY, Patel A, Sahoo DK. Harnessing the power of nutritional antioxidants against adrenal hormone imbalance-associated oxidative stress. Front Endocrinol (Lausanne). 2023 Nov 30;14:1271521. doi: 10.3389/fendo.2023.1271521. PMID: 38098868; PMCID: PMC10720671. www.ncbi.nlm.nih.gov/pmc/articles/PMC10720671/
- Aleman RS, Moncada M, Aryana KJ. Leaky gut and the ingredients that help treat it: a review. Molecules. 2023 Jan 7;28(2):619. doi: 10.3390/molecules28020619. PMID: 36677677; PMCID: PMC9862683. www.ncbi.nlm.nih.gov/pmc/articles/PMC9862683/
- National Research Council (US) Subcommittee on the Tenth Edition of the Recommended Dietary Allowances. Recommended Dietary Allowances: 10th Edition. Washington (DC): National Academies Press (US); 1989. 6, Protein and Amino Acids. Available from: www.ncbi.nlm.nih.gov/books/NBK234922/#
- de Souza RGM, Schincaglia RM, Pimentel GD, Mota JF. Nuts and human health outcomes: a systematic review. Nutrients. 2017;9(12):1311. Published 2017 Dec 2. doi:10.3390/nu91213112 https://pubmed.ncbi.nlm.nih.gov/29207471/
- Malaguarnera M, Cammalleri L, Gargante MP, et al. L-carnitine treatment reduces severity of physical and mental fatigue and increases cognitive functions in centenarians: a randomized and controlled clinical trial. Am J Clin Nutr. 2007;86(6):1738–1744. doi:10.1093/ajcn/86.5.1738
- Fairley JL, Zhang L, Glassford NJ, Bellomo R. Magnesium status and magnesium therapy in cardiac surgery: a systematic review and meta-

analysis focusing on arrhythmia prevention. J Crit Care. 2017;42:69–77. doi:10.1016/j.jcrc.2017.05.038

- Zhang X, Li Y, Del Gobbo LC, et al. Effects of magnesium supplementation on blood pressure: a meta-analysis of randomized double-blind placebo-controlled trials. Hypertension. 2016;68(2):324–333. doi:10.1161/HYPERTENSIONAHA.116.07664
- Ma F, Zhou X, Li Q, et al. Effects of folic acid and vitamin B12, alone and in combination on cognitive function and inflammatory factors in the elderly with mild cognitive impairment: a single-blind experimental design. Curr Alzheimer Res. 2019;16(7):622–632. doi:10.2174/156720501666619072514462 9
- de Jager CA, Oulhaj A, Jacoby R, et al. Cognitive and clinical outcomes of homocysteine-lowering B-vitamin treatment in mild cognitive impairment: a randomized controlled trial. Int J Geriatr Psychiatry. 2012;27(6):592–600. doi:10.1002/gps.2758
- Zhang YP, Lou Y, Hu J, et al. DHA supplementation improves cognitive function via enhancing ab-mediated autophagy in Chinese elderly with mild cognitive impairment: a randomised placebo-controlled trial. J Neurol Neurosurg Psychiatry. 2018;89(4):382–388. doi:10.1136/jnnp-2017-316176
- Zhang YP, Miao R, Li Q, et al. Effects of DHA supplementation on hippocampal volume and cognitive function in older adults with mild cognitive impairment: a 12-month randomized, double-blind, placebo-controlled trial. J Alzheimers Dis. 2017;55(2):497–507. doi:10.3233/JAD-160439
- Kucukgoncu S, Zhou E, Lucas KB, Tek C. Alpha-lipoic acid (ALA) as a supplementation for weight loss: results from a meta- analysis of randomized controlled trials. Obes Rev. 2017;18(5):594–601. doi:10.1111/obr.12528
- Travica N, Ried K, Sali A, et al. Plasma vitamin C concentrations and cognitive function: a cross-sectional study. Front Aging Neurosci. 2019;11:72. Published 2019 Apr 2. doi:10.3389/fnagi.2019.00072
- Coles LD, Tuite PJ, Öz G, et al. Repeated-dose oral n-acetylcysteine in Parkinson's disease: pharmacokinetics and effect on brain glutathione and oxidative stress. J Clin Pharmacol. 2018;58(2):158–167. doi:10.1002/jcph.1008
- Sanoobar M, Eghtesadi S, Azimi A, et al. Coenzyme Q10 supplementation ameliorates inflammatory markers in patients with multiple sclerosis: a double blind, placebo, controlled randomized clinical trial. Nutr Neurosci. 2015;18(4):169–176. doi:10.1179/1476830513Y.0000000106
- Sangsefidi ZS, Yaghoubi F, Hajiahmadi S, Hosseinzadeh M. The effect of coenzyme Q10 supplementation on oxidative stress: a systematic review and meta-analysis of randomized controlled clinical trials. Food Sci Nutr. 2020;8(4):1766–1776. Published 2020 Mar 19. doi:10.1002/fsn3.1492

- Nociti V, Romozzi M. The role of BDNF in multiple sclerosis neuroinflammation. Int J Mol Sci. 2023 May 8;24(9):8447. doi: 10.3390/ijms24098447. PMID: 37176155; PMCID: PMC10178984. www.ncbi.nlm.nih.gov/pmc/articles/PMC10178984/
- Shobeiri P, Karimi A, Momtazmanesh S, Teixeira AL, Teunissen CE, van Wegen EEH, et al. (2022) Exercise-induced increase in blood-based brain-derived neurotrophic factor (BDNF) in people with multiple sclerosis: a systematic review and meta-analysis of exercise intervention trials. PLoS ONE 17(3): e0264557. https://doi.org/10.1371/ journal.pone.0264557
- Poles J, Karhu E, McGill M, et al. The effects of twenty-four nutrients and phytonutrients on immune system function and inflammation: a narrative review. J Clin Transl Res. 2021;7(3):333–376.
- Luvián-Morales J, Varela-Castillo FO, Flores-Cisneros L, et al. Functional foods modulating inflammation and metabolism in chronic diseases: a systematic review. Crit Rev Food Sci Nutr. 2021;1–22. doi:10.1080/10408398.2021.1875189
- Kaur N, Chugh H, Sakharkar MK, Dhawan U, et al. Neuroinflammation mechanisms and phytotherapeutic intervention: a systematic review. ACS Chem Neurosci. 2020;11(22):3707–3731. doi:10.1021/acschemneuro.0c00427
- Hosseini B, Berthon BS, Saedisomeolia A, et al. Effects of fruit and vegetable consumption on inflammatory biomarkers and immune cell populations: a systematic literature review and meta-analysis. Am J Clin Nutr. 2018;108(1):136–155. doi:10.1093/ajcn/nqy082
- Hussain T, Tan B, Yin Y, et al. Oxidative stress and inflammation: what polyphenols can do for us? Oxid Med Cell Longev. 2016;2016:7432797. doi:10.1155/2016/7432797
- Schultz H, Ying GS, Dunaief JL, Dunaief DM. Rising plasma beta-carotene is associated with diminishing C-reactive protein in patients consuming a dark green leafy vegetable-rich, Low Inflammatory Foods Everyday (LIFE) diet. Am J Lifestyle Med 2019;15(6):634–643. doi:10.1177/1559827619894954
- Frugé AD, Smith KS, Riviere AJ, et al. A dietary intervention high in green leafy vegetables reduces oxidative DNA damage in adults at increased ris of colorectal cancer: biological outcomes of the randomized controlled meat and three greens (M3G) feasibility trial. Nutrients 2021;13(4):1220. doi:10.3390/nu13041220
- Al-Aubaidy HA, Dayan A, Deseo MA, et al. Twelve-week Mediterranean diet intervention increases citrus bioflavonoid levels and reduces inflammation in people with type 2 diabetes mellitus. Nutrients. 2021;13(4):1133. doi:10.3390/nu13041133

- SaeidiFard N, Djafarian K, Shab-Bidar S. Fermented foods and inflammation: a systematic review and meta-analysis of randomized controlled trials. Clin Nutr ESPEN 2020;35:30–39. doi:10.1016/j.clnesp.2019.10.010
- Mullins AP, Arjmandi BH. Health benefits of plant-based nutrition: focus on beans in cardiometabolic diseases. Nutrients. 2021 Feb 5;13(2):519. doi: 10.3390/nu13020519. PMID: 33562498; PMCID: PMC7915747. www.ncbi.nlm.nih.gov/pmc/articles/PMC7915747/
- Soliman GA. Intermittent fasting and time-restricted eating role in dietary interventions and precision nutrition. Front Public Health. 2022 Oct 28;10:1017254. doi: 10.3389/fpubh.2022.1017254. PMID: 36388372; PMCID: PMC9650338. www.ncbi.nlm.nih.gov/pmc/articles/PMC9650338/
- Parr EB, Devlin BL, Hawley JA. Perspective: time-restricted eating-integrating the what with the when. Adv Nutr. 2022 Jun 1;13(3):699–711. doi: 10.1093/advances/nmac015. PMID: 35170718; PMCID: PMC9156382. www.ncbi.nlm.nih.gov/pmc/articles/PMC9156382/
- Ratiner K, Shapiro H, Goldenberg K, Elinav E. Time-limited diets and the gut microbiota in cardiometabolic disease. J Diabetes. 2022 Jun;14(6):377–393. doi: 10.1111/1753-0407.13288. Epub 2022 Jun 13. PMID: 35698246; PMCID: PMC9366560. www.ncbi.nlm.nih.gov/pmc/articles/PMC9366560/
- Brandhorst S, Choi IY, Wei M, Cheng CW, Sedrakyan S, Navarrete G, Dubeau L, Yap LP, Park R, Vinciguerra M, Di Biase S, Mirzaei H, Mirisola MG, Childress P, Ji L, Groshen S, Penna F, Odetti P, Perin L, Conti PS, Ikeno Y, Kennedy BK, Cohen P, Morgan TE, Dorff TB, Longo VD. A periodic diet that mimics fasting promotes multi-system regeneration, enhanced cognitive performance, and healthspan. Cell Metab. 2015 Jul 7;22(1):86–99. doi: 10.1016/j.cmet.2015.05.012. Epub 2015 Jun 18. PMID: 26094889; PMCID: PMC4509734. www.ncbi.nlm.nih.gov/pmc/articles/PMC4509734/
- Yin Z, Klionsky DJ. Intermittent time-restricted feeding promotes longevity through circadian autophagy. Autophagy. 2022 Mar;18(3):471–472. doi: 10.1080/15548627.2022.2039524. Epub 2022 Feb 27. PMID: 35220894; PMCID: PMC9037462. www.ncbi.nlm.nih.gov/pmc/articles/PMC9037462/
- Vitall. UK statistics on vitamin & mineral deficiency 2023. 7 Mar 2023. https://vitall.co.uk/health-tests-blog/statistics-vitamin-mineral-deficiency-uk
- Forrest KY, Stuhldreher WL. Prevalence and correlates of vitamin D deficiency in US adults. Nutr Res. 2011 Jan;31(1):48–54. doi: 10.1016/j.nutres.2010.12.001. PMID: 21310306. https://pubmed.ncbi.nlm.nih.gov/21310306/

- Cleveland Clinic. Adaptogens. 02/10/2022. https://my.clevelandclinic.org/health/drugs/22361-adaptogens

Sleep

- Mental Health UK. Sleep and mental health. https://mentalhealth-uk.org/help-and-information/sleep/
- Nelson KL, Davis JE, Corbett CF. Sleep quality: an evolutionary concept analysis. Nurs Forum. 2022 Jan;57(1):144–151. doi: 10.1111/nuf.12659. Epub 2021 Oct 5. PMID: 34610163. https://pubmed.ncbi.nlm.nih.gov/34610163/
- Scott AJ, Webb TL, Martyn-St James M, Rowse G, Weich S. Improving sleep quality leads to better mental health: a meta-analysis of randomised controlled trials. Sleep Med Rev. 2021 Dec;60:101556. doi: 10.1016/j.smrv.2021.101556. Epub 2021 Sep 23. PMID: 34607184; PMCID: PMC8651630. https://pubmed.ncbi.nlm.nih.gov/34607184/
- Denison HJ, Jameson KA, Sayer AA, Patel HP, Edwards MH, Arora T, Dennison EM, Cooper C, Baird J. Poor sleep quality and physical performance in older adults. Sleep Health. 2021 Apr;7(2):205–211. doi: 10.1016/j.sleh.2020.10.002. Epub 2020 Nov 20. PMID: 33223446. https://pubmed.ncbi.nlm.nih.gov/33223446/
- Alanazi MT, Alanazi NT, Alfadeel MA, Bugis BA. Sleep deprivation and quality of life among uterine cancer survivors: systematic review. Support Care Cancer. 2022 Mar;30(3):2891–2900. doi: 10.1007/s00520-021-06589-9. Epub 2021 Sep 30. PMID: 34595604. https://pubmed.ncbi.nlm.nih.gov/34595604/
- Cruz T, García L, Álvarez MA, Manzanero AL. Sleep quality and memory function in healthy ageing. Neurologia (Engl Ed). 2022 Jan–Feb;37(1):31–37. doi: 10.1016/j.nrleng.2018.10.024. Epub 2021 Sep 10. PMID: 34518120. https://pubmed.ncbi.nlm.nih.gov/34518120/
- Smith PJ, Sherwood A, Avorgbedor F, Ingle KK, Kraus WE, Hinderliter AE, Blumenthal JA. Sleep quality, metabolic function, physical activity, and neurocognition among individuals with resistant hypertension. J Alzheimers Dis. 2023;93(3):995–1006. doi: 10.3233/JAD-230029. PMID: 37212110. https://pubmed.ncbi.nlm.nih.gov/37212110/
- Rosa JPP, Gentil P, Knechtle B, Vancini RL, Campos MH, Vieira CA, Andrade MS, de Lira CAB. Technology and sleep quality: friend or foe? Let the exergames come into play! Int J Sports Med. 2022 Aug;43(9):768–772. doi: 10.1055/a-1756-5005. Epub 2022 Mar 21. PMID: 35315004. https://pubmed.ncbi.nlm.nih.gov/35315004/
- Gilmour H, Lu D, Polsky JY. Sleep duration, sleep quality and obesity in the Canadian Armed Forces. Health Rep. 2023 May 17;34(5):3–14.

doi: 10.25318/82-003-x202300500001-eng. PMID: 37219888. https://pubmed.ncbi.nlm.nih.gov/37219888/

- Davis SM, Mekany M, Kim JJ, Han JJ. Patient sleep quality in acute inpatient rehabilitation. PM R. 2021 Dec;13(12):1385–1391. doi: 10.1002/pmrj.12550. Epub 2021 Feb 22. PMID: 33432699. https://pubmed.ncbi.nlm.nih.gov/33432699/
- Worley SL. The extraordinary importance of sleep: the detrimental effects of inadequate sleep on health and public safety drive an explosion of sleep research. P T. 2018 Dec;43(12):758–763. PMID: 30559589; PMCID: PMC6281147. www.ncbi.nlm.nih.gov/pmc/articles/PMC6281147/
- Zielinski MR, Systrom DM, Rose NR. Fatigue, sleep, and autoimmune and related disorders. Front Immunol. 2019 Aug 6;10:1827. doi: 10.3389/fimmu.2019.01827. PMID: 31447842; PMCID: PMC6691096. https://pubmed.ncbi.nlm.nih.gov/31447842/

Stress

- Wong SQ, Kumar AV, Mills J, Lapierre LR. Autophagy in aging and longevity. Hum Genet. 2020 Mar;139(3):277–290. doi: 10.1007/s00439-019-02031-7. Epub 2019 May 30. PMID: 31144030; PMCID: PMC6884674. https://pubmed.ncbi.nlm.nih.gov/31144030/
- Vikram A, Anish R, Kumar A, Tripathi DN, Kaundal RK. Oxidative stress and autophagy in metabolism and longevity. Oxid Med Cell Longev. 2017;2017:3451528. doi: 10.1155/2017/3451528. Epub 2017 Feb 28. PMID: 28337248; PMCID: PMC5350394. https://pubmed.ncbi.nlm.nih.gov/28337248/
- Galkin F, Kovalchuk O, Koldasbayeva D, Zhavoronkov A, Bischof E. Stress, diet, exercise: common environmental factors and their impact on epigenetic age. Ageing Res Rev. 2023 Jul;88:101956. doi: 10.1016/j.arr.2023.101956. Epub 2023 May 19. PMID: 37211319. https://pubmed.ncbi.nlm.nih.gov/37211319/
- Tang YY, Hölzel BK, Posner MI. The neuroscience of mindfulness meditation. Nat Rev Neurosci. 2015 Apr;16(4):213–225. doi: 10.1038/nrn3916. Epub 2015 Mar 18. PMID: 25783612. https://pubmed.ncbi.nlm.nih.gov/25783612/
- Basso JC, McHale A, Ende V, Oberlin DJ, Suzuki WA. Brief, daily meditation enhances attention, memory, mood, and emotional regulation in non-experienced meditators. Behav Brain Res. 2019 Jan 1;356:208–220. doi: 10.1016/j.bbr.2018.08.023. Epub 2018 Aug 25. PMID: 30153464. https://pubmed.ncbi.nlm.nih.gov/30153464/
- Orme-Johnson DW, Barnes VA. Effects of the transcendental meditation technique on trait anxiety: a meta-analysis of randomized controlled trials.

J Altern Complement Med. 2014 May;20(5):330–341. doi: 10.1089/acm.2013.0204. Epub 2013 Oct 9. PMID: 24107199. https://pubmed.ncbi.nlm.nih.gov/24107199/
- Chen KW, Berger CC, Manheimer E, Forde D, Magidson J, Dachman L, Lejuez CW. Meditative therapies for reducing anxiety: a systematic review and meta-analysis of randomized controlled trials. Depress Anxiety. 2012 Jul;29(7):545–562. doi: 10.1002/da.21964. Epub 2012 Jun 14. PMID: 22700446; PMCID: PMC3718554. https://pubmed.ncbi.nlm.nih.gov/22700446/

Strength

- Galkin F, Kovalchuk O, Koldasbayeva D, Zhavoronkov A, Bischof E. Stress, diet, exercise: common environmental factors and their impact on epigenetic age. Ageing Res Rev. 2023 Jul;88:101956. doi: 10.1016/j.arr.2023.101956. Epub 2023 May 19. PMID: 37211319. https://pubmed.ncbi.nlm.nih.gov/37211319/
- Ross A, Thomas S. The health benefits of yoga and exercise: a review of comparison studies. J Altern Complement Med. 2010 Jan;16(1):3–12. doi: 10.1089/acm.2009.0044. PMID: 20105062. https://pubmed.ncbi.nlm.nih.gov/20105062/
- Smith PJ, Merwin RM. The role of exercise in management of mental health disorders: an integrative review. Annu Rev Med. 2021 Jan 27;72:45–62. doi: 10.1146/annurev-med-060619-022943. Epub 2020 Nov 30. PMID: 33256493; PMCID: PMC8020774. https://pubmed.ncbi.nlm.nih.gov/33256493/
- Galloza J, Castillo B, Micheo W. Benefits of exercise in the older population. Phys Med Rehabil Clin N Am. 2017 Nov;28(4):659–669. doi: 10.1016/j.pmr.2017.06.001. PMID: 29031333. https://pubmed.ncbi.nlm.nih.gov/29031333/
- Brellenthin AG, Lanningham-Foster LM, Kohut ML, Li Y, Church TS, Blair SN, Lee DC. Comparison of the cardiovascular benefits of resistance, aerobic, and combined exercise (CardioRACE): rationale, design, and methods. Am Heart J. 2019 Nov;217:101–111. doi: 10.1016/j.ahj.2019.08.008. Epub 2019 Aug 15. PMID: 31520895; PMCID: PMC6861681. https://pubmed.ncbi.nlm.nih.gov/31520895/

Social

- Buettner D. The blue zones: secrets for living longer. National Geographic.
- McGlone F, Wessberg J, Olausson H. Discriminative and affective touch: sensing and feeling. Neuron. 2014 May 21;82(4):737–755. doi: 10.1016/j.

neuron.2014.05.001. PMID: 24853935. https://pubmed.ncbi.nlm.nih.gov/24853935/

- Xing Y, Zhang L, Zhang Y, He R. Relationship between social interaction and health of the floating elderly population in China: an analysis based on interaction type, mode and frequency. BMC Geriatr. 2023 Oct 16;23(1):662. doi: 10.1186/s12877-023-04386-z. PMID: 37845627; PMCID: PMC10580520. https://pubmed.ncbi.nlm.nih.gov/37845627/

Quality of food

- Bamford C. Nurture. Square Bag. 2018.
- Kiss the Ground. Netflix. 2020.
- Riverford. Why Organic. www.riverford.co.uk/ethics-and-ethos/why-organic
- Soil Association. What is organic food? www.soilassociation.org/take-action/organic-living/what-is-organic/

APPENDIX IV

Create your personalized plan

As you go through the book and find action steps that would benefit you or that you think you need to improve on write them out here and create your own personalized plan.

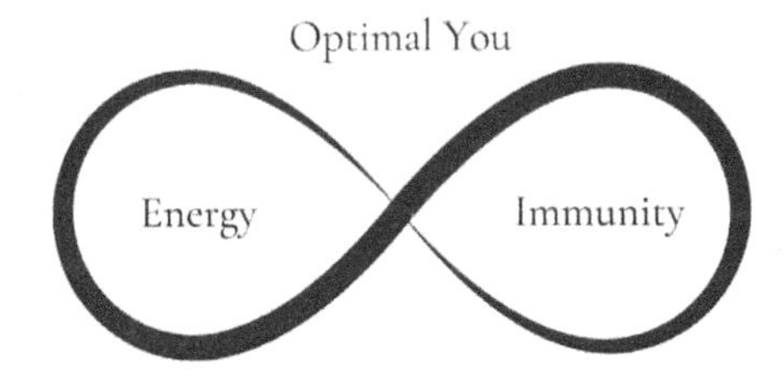

Create your personalized plan

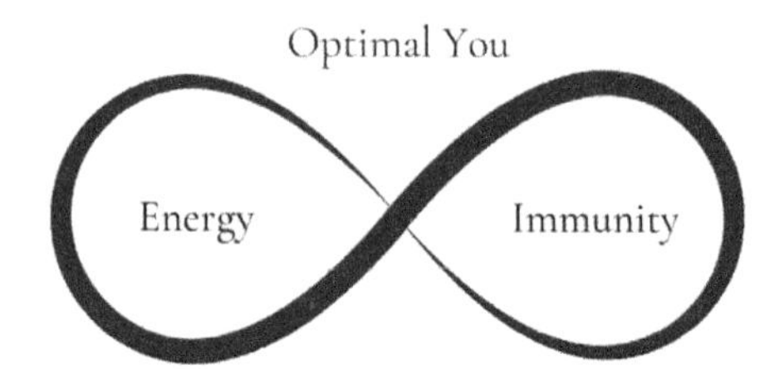

Create your personalized plan

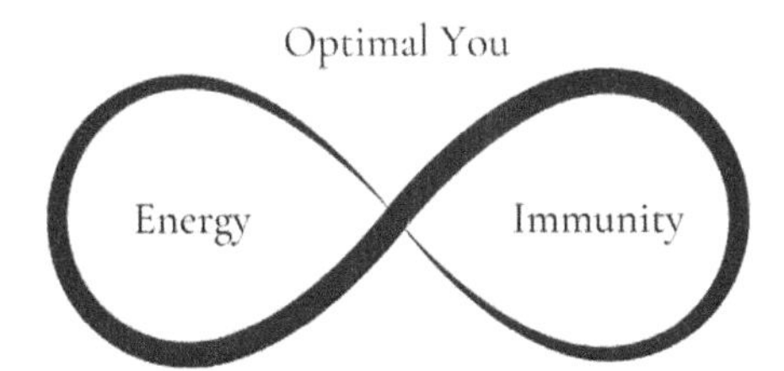

Create your personalized plan

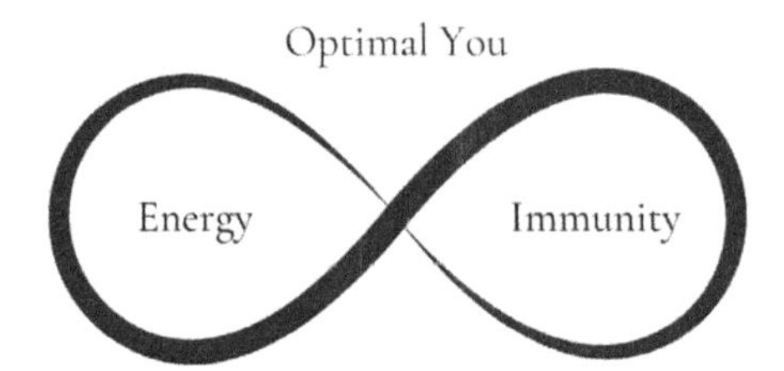

Acknowledgements

Writing a book is something I've always wanted to do. I'm an avid reader and I truly believe that books are a great way to reach and support more people. There are so many people that I could thank here who have helped me on this journey to becoming a published author (I still can't quite believe that as I write this it will actually be a reality soon). We set a pretty tight deadline to get the manuscript done and on top of that it has been quite a year for me so far, one that hasn't entirely gone to plan. Life has thrown some curve balls our way, but that only makes me believe in this book and the importance of optimal health even more. Alison Jones, Shell Cooper, Nim Moorthy, Michelle Charman, Frances Staton and all the team at PIP, thank you for your understanding and support through this journey. Also thanks to project manager Kelly Winter and editors Alana Clogan and Emily Boyd. You've all been a part of making this dream come true.

Next to my wonder team who made the cover photo shoot day a lot less daunting. Matt Priestly for capturing the front cover image. You saw my vision from our first phone call and made this 'awkward in front of a camera woman' feel at ease. Liz Clough for being a make up artist who got it when I said 'I don't wear much make up so I still want to feel like myself, just better than I can do myself.' Rebecca Kuk for decades of hair and friendship, always the best, always making me laugh.

To those who support me in my business. My mentor Chris Ducker for pushing me out of my comfort zone with the podcast and this book and for teaching me to be a 'Youpreneur'. Chloe Ducker for all your hard work editing the podcast, without you it definitely would not happen weekly! Hope Phillipson, my amazing creative assistant, I'm so glad I found you so early on in my business, you just understood my vision from day one. To Karen Kissane for creating business tools with epic support so we can actually get things done!

To those of you who took the time to read an early version of the manuscript: Emma Forbes, Dr Lafina Diamandis, Farzannah Nasser, Chris Ducker and my lovely patients; thank you for your lovely feedback and support.

To all of the guests who have agreed to come on the podcast and have honest conversations with me. The podcast was the start of this journey for me – it is what led to the book. When I started The Goode Health Podcast in 2023 I had no idea how it was going to go, but it is one of the favourite parts of my job. Being able to have these conversations with you all is an honour and in the same way as this book will, it has helped reach and support a wider and even more international audience. Many of you have become friends, but most importantly you have shared your stories and expertise with the world. I am always amazed every time someone contacts me to tell me how beneficial they found an episode. You helped make it possible.

To the functional testing labs both in the UK, USA and elsewhere, especially Cyrex, Doctors Data and Regenerus, for bringing pioneering tests to the forefront of healthcare. To the supplement companies and the people behind them who promote best forms, optimal dosages and sustainable practices. To Dr Datis Kharrazian for creating groundbreaking clinical training programmes for practitioners, learning from you is a privilege and an honour. To the Institute for Functional Medicine (and Clinical Education their UK partner), thank you for all the hard educational work you do – what an incredible job we have to be a part of this movement. To the functional medicine world as a whole, from the practitioners to the researchers, those training others, to those working in all different aspects of this amazing community we get to be a part of, you are the light at the end of the tunnel for many, the hope for the future.

From a personal perspective... Rachel Gaunt, my oldest friend, the sister I never had, for your support, love and endless encouragement. We grew up together; here's to growing old disgracefully together. Tom Elphick for years and decades of friendship and advice, you are a true friend, may we have many more. Polly Moore for endless insightful support and friendship, your wisdom and grace is inspiring. Eric Ho for helping me quieten the parrot and the owl and so many other friends I have met on this crazy journey of life.

Special mention to my parents, Andrew and Julie Goode, this book is dedicated to you, I wouldn't be who I am, or where I am today, without your support. For never pushing me, but always encouraging me, for supporting me when I couldn't work, to rooting for me now I can. I love you. To those who don't get to see this part of my journey, especially my grandparents; Joe, Molly, John and Elizabeth; Nana, Uncle John and Auntie Barbara, I wish you were here, I hope you are proud.

Finally to all my past, present and future clients, all over the world, for trusting in me, for taking the decision to work on your health, for listening to your own body and learning to advocate for yourself and for being bold and courageous enough to never accept anything less than optimal!

We all deserve to live optimally well.

About the author

Nicole Goode is a Certified Functional Medicine Health Consultant (CFMHC), Registered Nutritional Therapist (BANT CNHC) and BANT-registered Nutritionist. An associate of the Royal Society of Medicine and a member of The Institute for Functional Medicine, she is the founder and Clinical Director of The Goode Health Clinic, named Best Functional Medicine Clinic UK for two years running in 2023 and 2024 at the Health, Beauty and Wellness Awards. Podcaster and host of The Goode Health Podcast, Nicole has also contributed to publications such as Women's Health and Marie Claire. This is her first book.

Index

Entries beginning with a number are indexed as spelled out; e.g. 5-methyltetrahydrofolate is listed under F. Locators in *italics* indicate figures or tables. Entries beginning with The are indexed under the second word, e.g. The MitoImmune Method is listed under M.

A

adaptive immune systems 48–49, *49*
adaptogens 158–159
adenosine 135
adenosine triphosphate (ATP) 9, 38, 40, 41, 43, 45
adrenal dysfunction, symptoms of 72
adrenal fatigue 73
adrenal health 34, 164–165
adrenal nutrients *76*
adrenal testing 196–197
adrenal-brain connection 70–71
adrenal-immune connection 52, 70
adrenaline (epinephrine) 52, 71, 135
adrenal-mitochondria connection 42–43, 69–70
adrenals, the 64, 67, 100
  and hormones 71
  and the immune system 52, 70
  understanding 69–72
adrenal-thyroid connection 69–72, 79
ageing 10, 50, 99
agriculture, local 185
Air Doctor 188
air pollutants 191
air purifiers 188
alcohol, reducing intake 133–134
alpha-lipoic acid (ALA) 139–140
Alzheimer's Association 27
Alzheimer's disease 10, 27, 39, 41, 61, 135
amino acids 62
amygdala 64
antecedents 26
antibiotics 182–183, 184
antibodies 51, 82, 83, 84, 85–86
anti-inflammatory foods 129–130
artificial fragrances and dyes 192–193
Ashwagandha (supplement) 159
autoantibody testing 198
autoimmune disease 6–8, 41, 48 and note
  and stress 7
  thyroid conditions 81–85
autoimmune spectrum 48, 54–55
autoimmune vs immune 48
automatic nervous system 67–68
autophagy 136

B

B vitamins 45, 62, 103, 140–141, *141*
  see also individual vitamins
bacteriophages 94
balanced plate guide 128
BANT (British Association for Nutrition and Lifestyle Medicine) 156
Bare Biology 158

berberine 148–149
betaine 104
bisphenol A (BPAs) 190
blood-sugar balance 109–111
blood-sugar imbalance, symptoms of 111
blue light vs natural light 163
blue zone communities 173 and note 175
Bodybio 160
body scan 170
brain, the
  and the adrenals 70–71
  and the gut 89–90
  and the immune system 52
  and the thyroid 79
brain atrophy 61
brain function 62–63, 65
brain health
  case study 57–58
  defined 34
  and environment 64
  and lifestyle 63–65
  maintaining 57–59
  physical health link 59
  and sense of purpose 64
  and sleep 63–65, 163
  and social connections 64
brain nutrients *65*
brain reserve vs cognitive reserve 61–62
brain-derived neurotropic factor (BDNF) 62, 140
brain-immune axis 52
Breath In Breath Out (Sandeman) 169
breath work 169, 170
breathing 102, 137
Buettener, Dan 173
burnout 8, 41, 72
butyrate 89

C

caffeine, reducing intake 134–135
candida 93
cardiometabolic disorders (CDMs) 108–109
cardiometabolic function, role of thyroid in 80
cardiometabolic health
  around the body 113
  balance and optimizing 107–108
  defined 34
  and immune system 53
  and mitochondria 44
  and sleep 165
cardiometabolic nutrients *115*
cardiovascular disease (CVD) 107, 108, 112–113, 134
cardiovascular health 107
carnitine 142
carotenoids 142
casein 83
cast iron cookware 190
cause vs symptoms approach 16
cell death 39, 61
cellular energy, defined 9
cellular health 18, 78
ceramic cookware 191
Chatterjee, Rangan 175
chemical goitrogens 84
chewing 136–137
cholesterol 42, 71, 111–113
chlorine 104, 143
chronic care vs emergency care 15–16
chronic disease *4*
chronic fatigue 31
chronic health conditions 17
chronic inflammation 47–48, 53
chronic stress 45, 54, 79, 114, 199
  and cortisol curve 73
circadian rhythm 45, 73, 135, 136, 163
cobalamin (vitamin B12) 62, 103, 140–141
coeliac disease 82, 90
coenzyme Q10 (CoQ10) 143–144
cognitive reserve vs brain reserve 61–62
community 177–179
conventional approach vs functional medicine 15–18
cortisol 43, 64, 69, 99, 102
  and neurotransmitters 70–71
cortisol curve 73–74, 195
'cortisol steal' 71, 72

curcumin 149
cytokines 52, 53, 64, 79, 99

D
dairy 54, 82–83
dedication vs fast approach 17
dehydroepiandrosterone (DHEA) 71
dementia 39
Designs for Health 158
detoxification pathways 100–102, *102*
diabetes 109, 111
diet, poor 33
digestion 136
disease centred approach vs patient centred approach 16
The Diversity Checklist 126
docosahexaenoic acid (DHA) 62
dopamine 71, 98
dysbiosis 84, 92
dyslipidemia 108–109

E
eating behaviours
  breakfast like a king 136
  eating on the run 136–137
  intermittent fasting 135–136
  seasonal eating 184–185
Edison, Thomas A. 18
eicosapentaenoic acid (EPA) 62
elevation, and walking 175
elimination diet 96
emergency care vs chronic care 15–16
endocrine disruptors 190–191
endocrine glands, roles of 99–100
endocrine system 77, 79, 90, 99
endogenous antioxidants 39
endurance 40
energy 5, 31, 40, 44
  defined 38
  and the thyroid 78
enteric nervous system (ENS) 89–90
enteroendocrine cells 90
environment 27, 45, 64
environmental toxins 121
  air pollutants 191
  beauty products 191–192
  endocrine disruptors 190–191
  heavy metals 188–189
  mould 187–188
  and the thyroid 80–81
epigenetics 18
Erasmus, Desiderius 18–19
eukaryotic viruses 94
exercise
  benefits of 172
  and brain function 65
  see also physical activity
exogenous antioxidants 39

F
farm-to-table foods 185
fast vs dedication approach 17
fatigue 40, 41
fibre 122–123
fight or flight response 31, 67–68, 69, 71, 75
fish 189
  5-methyltetrahydrofolate (5-MTHF) 103
flavonoids 144
flaxseed 130
folate (B9) 103
follicular cell destruction 83
food goitrogens 83–84
foods
  anti-inflammatory 129–130
  chewing 136–137
  farm-to-table 185
  genetically modified 186
  plant foods 123, 127–128
  quality 121
formaldehyde 192
free radicals 38–39, 45, 139
full blood panel 196
full thyroid panel 196
functional medicine
  vs conventional approach 15–18
  defined 15
  timeline 19–21, *21*
  tree *23*

functional medicine practitioner, working with 200
functional testing 28, 121

G
GABA (Gamma-Aminobutyric Acid) 70
gastrointestinal (GI) tract 44
 see also gut health
genetically modified foods 184
glass cookware 190
gliosis 61
glucagon 105
glucose 44, 97, 109–110
glutamate 39, 54
glutamine 145
glutathione 39, 54, 145–146, 159
gluten 54, 82, 84, 91
 reduce or remove 131–132
glymphatic system 63
glyphosate 182
goitrogens 83–84
gratitude journalling 170
Grave's disease *81*
growth hormone (GH) 98, 105
gut health *92*
 defined 34
 the 5 Rs *95*
 food as trigger 95–96
 and the immune system 53, 88–89
 importance of 88–90
 infections and toxicities 92–95
 and mitochondria 44, 89
 and sleep 165
 testing 195
gut microbiome 87–88, 90–91, 182
gut nutrients *96*
gut-associated lymphoid tissue (GALT) 53
gut-brain axis 89–90
gut-endocrine connection 90
gut-immune axis 53, 88–89
gut-mitochondria connection 44, 89
gut-thyroid connection 80, 90

H
Hashimoto's disease 7, 79, *81*
 case study 43–44, 84–85
 and iodine 83
Haynes, Anthony 89
HDL cholesterol 111–112
health history 19–21, *19*, *33*
health journey 11–12, *12*, 200
healthspan
 defined 4
 extending 200
 vs lifespan *4-5*
heavy metals 118–189, 197
herbicides 182, 190
herbs and spices 125, 130
HLA-DQ 82
hobbies 171
holistic wellness *4*
hormonal balance, importance of 98–99
hormone health
 case study 99
 defined 34
 detoxification pathways 100–102
 and sleep 165
 testing 197
hormone nutrients *106*
hormones 105
 and the adrenals 71
 cognitive function and optimal health 105
 and the immune system 53
 and mitochondria 44
 overview 97–98
 and the thyroid 80
Huntingdon's disease 61
hydration 54, 130
hydrogenated fats 133
hyperglycaemia 111
hyperthyroidism 80, 81–82, *81*
hypothalamic-pituitary-adrenal (HPA) axis 67, 68, 70–71, 73, 77, 79
 dysfunction, stages of 74–75, *75*
hypothalamus 100
hypothyroidism 7, 80, 81–82, *81*

I
immortality 3
immune function, and sleep 163
immune health
 defined 34
 and inflammation 47–48
 nutrition and lifestyle 53–54
immune nutrients *54*
immune resilience
 defined 9–10
 focus on 47
immune response 88–89, 94
immune system 5, 91
 and the adrenals 52, 70
 and alcohol 133–134
 and the brain 52
 and cardiometabolic health 53
 and gut health 53, 88–89
 and hormones 53
 innate and adaptive 48–49, *49*
 and mitochondria 41, 52
 strong but balanced, need for 49–50
 and the thyroid 79
 and thyroid health 52
immune vs autoimmune 48
immunity 5, 31
immunosenescence 50
impaired glucose metabolism 109
inflammation 42, 43, 44, 47, 55, 82, 91
 and brain health 61
 and immune health 47–49
innate immune system 48–49, *49*
insoluble fibre 123
Institute for Functional Medicine 169
insulin 44, 97, 105
 resistance 109, 111
intestinal permeability 44, 54, 70, 80, 82, 91
iodine 83–84
ION (Institute for Optimum Nutrition) 155
iron 53, 146–147
Islets of Langerhans 100

J
journalling 170–171

K
ketosis 136

L
LDL cholesterol 71, 111
lectins 55, 84–85
legumes and pulses 124
leptin 105
lifespan vs healthspan *4-5*
lifestyle 27, 121, 139, 199
 and brain health 63–65
 and metabolic disorder 114
 and mitochondria 45–46
lifestyle: the 4S's
 sleep 161–169
 social 177–180
 strength 172–177
 stress 169–172
lipid profile 111–113
liver, the 100–101
 detoxification pathways *102*
longevity 3, 99
longevity medicine, defined 4

M
magnesium 147
mediators 26
meditation 170
melatonin 73, 98, 163
mental clarity 40
metabolic disorder 114
metabolic syndrome 108
metabolism 40, 78, 113, 136
methionine 103, 104
methylation 102–104
methylenetetrahydrofolate reductase (MTHFR) 103
microbiota 87
mito *see* mitochondria
mitochondria 8, 78
 and the adrenals 42–43, 69–70
 and cardiometabolic health 44

explained 37
and gut health 44, 89
and hormones 44
and the immune system 41, 52
impact on health 40
and lifestyle 45–46
and nutrition 45
mitochondrian nutrients *46*
mitochondria-brain connection 41–42
mitochondrial DNA (mtDNA) 37, 41, 78
mitochondria-thyroid connection 43–44
The MitoImmune Health Assessment 13, 34, 121, 139, 199
The MitoImmune Method 8, *31*
author's experience 31–34
pillars of *31*, 34
The MitoImmune Nutrition Plan
action step 1: phytonutrients 126
action step 2: fruit and vegetable balance 126–127
action step 3: 50 foods 127–128
action step 4: balanced plate guide 128
action step 5: balance your proteins 128–129
action step 6: anti-inflammatory foods 129
action step 7: herbs and spices 130
action step 8: flaxseed 130
action step 9: hydrate 131
action step 10: reduce or remove gluten 131–132
action step 11: reduce sugar 132–133
action step 12: reduce processed foods 133
action step 13: reduce alcohol 133–134
action step 14: reduce caffeine 134–135
filtered water 125
healthy fats 123–124
herbs and spices 125
legumes and pulses 124
nuts and seeds 125
plant foods 123
whole foods 122–123
The MitoImmune Plan 82
The MitoImmune Programme 114, 119–122, *120*
eating behaviours 135–137
implementing 120–122
molecular mimicry 82–83
mood 40
mould *see* mycotoxins
multiple sclerosis (MS) 39
mycotoxins 94, 178–188
testing 197

N

N-acetyl-cysteine (NAC) 145–146, 159
neurodegeneration 61
neurogenesis 65
neurological disease 60–61, *60*
neuronal injury 61
neuroplasticity 59–60, *59*, 62
neurotoxic substances 64
nightshades 55, 85
nutraceuticals *see* supplements
nutrients 121
and brain function 62–63
deficiencies 84
phytonutrients 62, 91, 102, 123
see also supplements
nutrigenomic DNA testing 198
nutrition 27
and autoimmunity 54–55
and immune health 53–54
and metabolic disorder 114
and mitochondria 45
and seasonal eating 184–185
nutritional status, testing 197
nuts and seeds 125, 134

O

Ochratoxin A 41, 187–188
octinoxate 193
oestrogen 98, 99, 105

omega 3 fatty acids 48, 54, 62, 123, 158, 189
optimal health *33*
  defined 4
  importance of 10–11
optimizing your health, defined 4
organic farming 184
organic foods 181–184
  checklist 183
  what it means to buy 182
organic wine 134
ovaries 97
oxidative stress 38–39, 43, 54, 69, 139
oxybenzone 193
oxygen 39

P
Panax Ginseng (supplement) 159
pancreas 97, 100
parabens 190, 192
parasites (gut) 92–93
parasympathetic nervous system 137
parasympathetic state 68
parathyroid glands 100
Parkinson's disease 39, 41, 61, 135
partially hydrogenated fats 133
People Pleasers 179
personalized care/medicine *4*, 13, 122
  importance of 29
  power of 27–29
  and sleep 165
  and testing 195
personalized plan templates 225–228
pesticides 64, 134, 157, 181, 182, 178, 190
petroleum-based ingredients 193
phospholipids 150–151, 159
phosphatidylcholine (PC) 150–151, 159
phosphatidylserine (PS) 150–151
phthalates 190, 192
physical activity 46, 54, 114, 136
  see also exercise
physical health, link to brain health 59
phytonutrients 62, 91, 102, 123
Pilates 173–174
pituitary 83, 97, 100
plant foods 123, 127–128
plastics 190–192
polyethylene glycols (PEGs) 192
polyphenols 134, 148–150
polyreactive autoimmunity, case study 94–95
poor health, cost of 7, 11, 17–18, *72*
power nap 168–169
prebiotics 151
pre-diabetes 109
prefrontal cortex 64
pregnenolone 42
preventative approach 18–19
  vs reactive approach 17
proactive health management *5*
probiotics 151, 158
processes foods, reducing intake 133
progesterone 98, 105
progressive muscle relaxation 170
proteins 124
  balancing 129–129
  substitutions 129
purpose, sense of and brain health 64
pyridoxine (vitamin B6) 62, 104, 141
pyrroloquinoline Quinone (PQQ) 152

Q
quality of life 104
quercetin 150

R
reactive approach vs preventative approach 17
reactive oxygen species (ROS) 39, 41, 78
recovery 40
red wine 40, 134
resveratrol 40, 134, 149–150
retirement 177
Rhodiola (supplement) 159
riboflavin (vitamin B2) 103, 140, 141

root cause medicine 23–26
antecedents, triggers and mediators 26
multiple sclerosis and *25*
one cause; many conditions 25–26, *25*
one condition; many causes 24, *24*
and testing 195
Roundup 182

S

saliva 136
SAMe 104
Sandeman, Stuart 169
screens 163, 166, 167
seasonal eating 184–185
secretory immunoglobulin A (sIgA) 88
selenium 152–153
serotonin 70, 98
sex hormones
and the adrenals 71
balance and optimizing 104
short-chain fatty acids (SCFAs) 88, 89, 94
siesta, the 168–169
The 6 Colours of Diversity 126
sleep
and brain health 63–65, 163
benefits of 163–164
caffeine and 168
calm mind and 167
and cardiometabolic health 165
checklist 168
exercise and 168
and gut health 165
and hormone health 165
and immune function 163
and lifestyle 161–169
morning routing 166–167
natural light vs blue light 163
non-rapid eye movement (NREM) 162
other steps 167–168
and personalized care 165
rapid eye movement (REM) 162, 169
routine, implementing 166
siesta 168–169
sleep cycle *162*
sleeping well 165–166
stages of 162
working time and 167
social (lifestyle S's)
build your community 177–179
learn to say NO 179
purpose in life 179–180
taking a day off 180
toxic behaviour 178
values 180
social connections, and brain health 64
sodium (salt) 55, 85
Sodium Laureth Sulfate (SLES) 192
Sodium Lauryl Sulfate (SLS) 192
soil health 159, 181
soluble fibre 123
spoon theory 176–177
sport 173
stainless steel cookware 190
stool test 196
strength (lifestyle S's)
and retirement 177
simple steps 176–177
3 P's 176–177
stress 11, 31, 33, 69, *69*
and autoimmune disease 7
and brain health 63
breath work 169, 170
checklist 171
and cortisol 71
cost of 72
defined 68–69
and journalling 170–171
lifestyle (the 4S's) 169–172
and meditation 170
and the nervous system 67–68
and relaxing hobby 171
support 171
see also chronic stress
stress hormone 42, 63–64
sugar

  impact on health 132
  ones to avoid *132*
  reducing intake 132–133
supplements 156–160
suprachiasmatic nucleus (SCN) 163
sweating 102
symptoms vs cause approach 16
system-based approach vs whole body approach 16

T

telomeres 18
testes 97
testosterone 98, 99, 105
TGabs 82
3-4-5 breathing 137
thymic atrophy 79
thyroid, the 77–78, 97, 100
  and the adrenals 69–72, 79
  balancing 78
  and the brain 79
  cardiometabolic function, role in 80
  and energy 78
  and environmental toxins 80–81
  and the gut 80, 90
  and the immune system 79
  and other hormones 80
  and mitochondria 43–44
  and sleep 165
thyroid conditions 7–8, *8*, 81–82
thyroid health, defined 34
thyroid hormones 43, 72, 78–80, 81, 83, 85, 195
thyroid nutrients *86*
thyroid releasing hormone (TRH) 77
thyroid stimulating hormone (TSH) 77, 78, 82, 83, 85, 195
thyroid testing 85–86
thyroxine (T4) 43, 77, 78, 82, 85
tight junction (intestine) 44, 70, 80, 91
time-restricted eating 135
T-lymphocytes 49, 79
toluene 193
TOPAbs 82, 84
toxic behaviour 178
trans fats 133
triclocarban 192
triclosan 192
triggers 26
triglycerides 112
triiodothyronine (T3) 43, 77, 78, 82, 85
type 1 deiodinase 43
type 2 deiodinase 43

V

vagus nerve 90
Vim & Vigour 158
viral panel 197
virome 93–94
vitamin A 153
vitamin B2 (riboflavin) 103, 140, 141
vitamin B6 (pyridoxine) 62, 104, 141
vitamin B12 (cobalamin) 62, 103, 140–141
vitamin C 39, 53, 62, 154, 159
vitamin D 38, 53, 82, 111, 124, 154–155, 157–158
vitamin E 39, 53, 62, 155

W

walking 174–175
water 130
  filtered 125
water filtration systems 188
wellness medicine, defined 4
what is health approach 16
what vs why approach 16, 24, *25*, 28
wheat germ agglutinin 84
whole body approach vs system-based approach 16
whole foods 122–123
why vs what approach 16, 24, *25*, 28

Y

yeast 93
yoga 173

Z

zinc 155–156

A quick word from Practical Inspiration Publishing...

We hope you found this book both practical and inspiring – that's what we aim for with every book we publish.

We publish titles on topics ranging from leadership, entrepreneurship, HR and marketing to self-development and wellbeing.

Find details of all our books at: www.practicalinspiration.com

Did you know...

We can offer discounts on bulk sales of all our titles – ideal if you want to use them for training purposes, corporate giveaways or simply because you feel these ideas deserve to be shared with your network.

We can even produce bespoke versions of our books, for example with your organization's logo and/or a tailored foreword.

To discuss further, contact us on info@practicalinspiration.com.

Got an idea for a business book?

We may be able to help. Find out more about publishing in partnership with us at: bit.ly/PIpublishing.

Follow us on social media...

@PIPTalking

@pip_talking

@practicalinspiration

@piptalking

Practical Inspiration Publishing

www.ingramcontent.com/pod-product-compliance
Lightning Source LLC
Jackson TN
JSHW061340131224
75386JS00008B/393